Praise for

It's the Little Things

"In her socially penetrating new book, Williams offers an often hilarious catalog of those painful encounters as they occur in public places, at home, at work, and in the media.... The result is a book that should be put in every schoolchild's hand as soon as the youngster can understand it."

—*The Boston Globe*

"Sassy and informative, *It's the Little Things* lets blacks and whites walk a mile in each others' shoes. The book's revelations won't erase the color line but might help sensitize blacks and whites about how their actions are viewed by others." —*The Christian Science Monitor*

"*It's the Little Things* is a lighter discourse on the ultra-serious matter of race in America.... Crafted to keep readers engaged, and also to avoid the angry shrieking that has silenced so much attempted dialogue between blacks and whites... A sounding board for blacks and whites concerned with bridging the racial divide." —*Newsday*

"No one could disagree with [Williams's] conclusion that 'what the races need more than anything else is to lighten up,' to cut one another some slack, think before they act on questionable presumptions about other people and memorize a few hitherto unsuspected offenses to be avoided."

—*The New York Times Book Review*

"[*It's the Little Things* is] the kind of reading that will make some black folks chuckle and nod their heads in appreciation." —*African Sun Times*

"Ms. Williams aims to chronicle 'the racial slights and indignities' that occur between the races—the everyday kinds of 'little things' that wouldn't get mentioned in more formal racial discussions sponsored by government or academia."
—*The New York Times*

"Based on Williams' own experiences and her work with focus groups, this book examines the leaps that must be made to bring about a greater level of trust between races."
—*The New York Daily News*

"With a wit that takes the sting out of sensitive subjects, Williams, who is black, examines issues rarely touched upon in interracial settings in the hopes of helping form a foundation for an improved conversation on race.... Williams helps us understand that whether or not there is a scientific basis for racial differences, in this country race really does play a large role in shaping our perceptions of ourselves, others, and issues and events around us."
—*The American Prospect*

"An excellent book, Williams presents a clear, honest yet humorous picture of the little things that often interfere with communication and friendship between blacks and whites. The examples she uses are pointed and eye opening. Reading *It's the Little Things* is a must for those of us who are working to improve relationships and understanding across racial lines."

—Alvin Poussaint, M.D., Professor of Psychiatry, Harvard Medical School

"Lena Williams soundly awakens us to the true importance of words. She clearly documents that hurtful words are the foundations of prejudicial and racist attitudes, behaviors, and actions. How wonderful it is to be so well informed after reading this book." —Camille O. Cosby, Ed.D.

"William's provocative book is sure to stimulate much discussion with its candid depiction of race relations."

—*Publishers Weekly*

"Williams, a reporter for the *New York Times,* speaks from experience about a range of annoying to dangerous incidences that are caused by the lack of understanding between the races. . . . Revealing, sometimes amusing, look at the sad state of race relations." —*Booklist*

It's the Little Things

IT'S THE LITTLE THINGS

Everyday Interactions That Anger, Annoy, and Divide the Races

LENA WILLIAMS

A Harvest Book • Harcourt, Inc.
San Diego New York London

Requests for permission to make copies of
any part of the work should be mailed to the
following address: Permissions Department,
Harcourt, Inc., 6277 Sea Harbor Drive,
Orlando, Florida 32887-6777.

www.HarcourtBooks.com

Library of Congress Cataloging-in-
Publication Data
Williams, Lena.
It's the little things: the everyday interactions
that get under the skin of Blacks and
Whites/Lena Williams.—1st ed.
p. cm.
ISBN 0-15-601348-7
1. United States—Race relations—
Psychological aspects. 2. Social
interactions—United States. 3. Afro-
Americans—Psychology. 4. Whites—United
States—Psychology. I. Title
E185.615.W4925 2000
305.8'00973—dc21 00-037027

Text set in Sabon
Designed by Leslie Fitch

Printed in the United States of America
First Harvest edition 2002
K J I H G F E D

To my late mother,
Lena Williams,
for giving me the only
thing that really mattered—
unconditional love

CONTENTS

Foreword

Over the many years that I've known Lena Williams—as a graduate student/baby-sitter, always happy to oblige so she could watch the Knicks games on my cable TV and peruse my books on journalism and as a novice reporter, later blossoming into a first-rate, seasoned veteran and peer—I have rarely seen her angry. But when her temperature does start to rise, Lena has a way of letting people know it. "That is about to get on my last nerve," she says and the one responsible knows better than to push it.

This came to mind as I began reading *It's the Little Things,* a book that strikes powerfully at the things that get on "my last nerve." Like when I first moved to South Africa. Apartheid had ended, to be sure. But it had only ended three years earlier, and I suspected from my experiences growing up in America's segregated South that apartheid would take a little longer to end in the heads and hearts of many white South Africans. So I went, not with a chip on my shoulder, but with my racial antennae finely tuned and ready to detect any hint of residual apartheid. Thus, when I had my first encounter with a white person who read my name off of an application and then addressed me as Charlayne, I quickly told her that until such time as we became friends, which was not necessarily outside the realm of possibility, I would appreciate it if she would give me the respect of calling me by my surname and the appropriate honorific.

This happened on numerous other occasions, prompting my show-me-some-respect mantra. Meanwhile, I was having the same experience with my black house staff, whom I addressed, as I had done in the states, by honorifics. White South Africans, of course, had always referred to blacks (with the possible exception of Nelson Mandela and Desmond Tutu) by their first names. One morning, I heard my housekeeper on the phone with a merchant. Four or five times, I heard her repeat, "You can call me Mrs. K." She was Mrs. K because we had removed the excuse whites made for not calling blacks Mr. or Mrs./Miss—"I just can't pronounce those African names"—and instructed Mrs. Kgoadi and my eighty-two-year-old mother's caregivers to use their initials. (We have Mrs. K, Miss P, and Miss T.) But the white woman on the other end of the phone was determined to have Mrs. K's first name. What's up with that?—as Lena would say. I took the receiver from Mrs. K's hand and identified myself, demanding to know what was it about Mrs. K's explanation that the woman didn't understand. "Er, Er, Well, it's just that, well..." I listened as she stumbled through an incomprehensible explanation and then told her that I would be handing the phone back to Mrs. K and requesting that she be addressed by her proper name.

That was not a benign instance. But, on the other hand, what I came to appreciate about South African society, is that there is a decided informality to it, including the fact that most South Africans rarely get really dressed up for anything. They are quick to identify themselves by their first names, although I have rarely seen this done reciprocally with black domestics and other household help. Still, after two years, I get a little less agitated when I'm called by my first name, though I quickly say, "Now what is your first name?"

It really IS the little things.

Before I moved to South Africa, I was once having my make-up done before going on the air to do a discussion segment with several guests on PBS's *The NewsHour with Jim Lehrer.* An older white man, a guest on the segment, was also in the dressing room. I spoke to him, but he barely looked in my direction, let alone comprehended my telling him that I was the anchor doing the segment. On the set, I got his attention with a series of well-informed questions. I could see a transformation unfolding right before my very eyes. After the program was over, he followed me back to the dressing room and lingered after his own make-up was removed and the other guests had left. I could sense that he was trying to work through something but that he didn't quite know how. I decided not to intrude. Finally, he spoke. "How long have you been doing this kind of work?" he asked.

"Oh, about twenty years," I answered.

He paused as that sank in, then said, "Well, I guess it beats being a chorus girl!"

I smiled and said, "I guess it does."

It really IS the little things.

Still, I could take this otherwise insensitive remark lightly because I felt it really was benign, an awkward, inexperienced attempt at a compliment. Not like the time when I was working at my alma mater, Lena's current employer, the *New York Times,* a great experience for me, as it has been for Lena—mostly. Except for little things, such as the occasion when a group of black reporters was working late and had drifted over to the part of the newsroom where the daily papers were kept. As we were exchanging little bullshit kinds of things, a white magazine editor from the ninth floor approached us.

"Having a meeting?" he asked cheerily.

"Yes," I answered, adding, "But don't worry, we've voted to keep you."

Not really sure of exactly what I had meant, but clearly deciding to go with the jovial flow of the moment, he said, "Oh, that's good."

One of us then asked him what he was doing in the newsroom.

"Oh," he said, "I came down here looking for writers for the magazine, but I see nobody's here."

As the other blacks looked at him in disbelief, I quickly chimed in: "I think we just changed our earlier vote."

He departed, looking a bit puzzled.

Fast forward... back to South Africa. On my way in

from the airport in Johannesburg one day, I was talking with the black driver about life in post-apartheid South Africa, where for the first time, the country's constitution, written by both blacks and whites, guarantees him rights that he never before had, including the right to dignity. But, he says during the course of the conversation, "Being black is still a twenty-four-hour-a-day job." I suspect that's true, in large part, because of the little things.

W. E. B. Dubois was prescient when at the turn of this last century he argued that the issue of the twentieth century would be that of the color line. That the problem lingers into the twenty-first century is not a comment on DuBois's lack of foresight, but rather on our failure to acknowledge his prediction and take it seriously. Lena Williams has done that in a most creative way—with perception and with a humor that takes some of the edge off the still simmering rage brought on by having the job of being black twenty-four hours a day. Reading her book won't solve the race problem—for it, like ignorance, may be ever with us. But from the South of America to South Africa, and all places in between, where laws require equality and a respect for human dignity, being mindful of the "little things" is the new frontier.

Lena Williams has given blacks, whites, and people of all hues a roadmap that can help us navigate a different racial reality in the twenty-first century.

Charlayne Hunter-Gault
June 1, 2000, Johannesburg

It's the Little Things

Introduction

I never liked the circus. Dancing and defecating animals; two-faced clowns; acts of derring-do, we blacks used to describe it as "white folks' foolishness."

The first time I set foot inside a circus tent was in 1955. I was five years old, and my kindergarten class went to see the Barnum & Bailey Circus at the Armory in Washington, D.C.

Back then, the nation's capital, like the rest of the nation, was segregated—if not by law, by design. Negroes—as we were known—were trying to prove to whites that we were worthy to walk among them, so we did little things like never going to a public event, be it the circus or church, without being dressed in our Sunday finest. There was no loud talking or misbehaving lest we "mark the race."

Our class was seated near the center ring, in the front rows. We were flanked by white students dressed casually in worn jeans, corduroy pants, or simple cotton dresses. A white-faced clown appeared and began handing out balloons to the white kids seated to our left. Naturally we wanted balloons, but we sat like perfect little ladies and gentlemen, patiently waiting our turn. When the clown reached our section, he took one look at our little black faces and kept walking. He then proceeded to the white children seated on our right and handed out more balloons.

"How come the clown didn't give us balloons?" we asked our teacher.

She told us to be quiet and promised to buy us balloons after the circus. She did, and we left happy.

Although I was a child, somehow I knew, then and there, that it was my brown skin the clown reacted to. If only I had been white I, too, would have gotten one of the clown's balloons, which from my child's eye seemed so much bigger and brighter than those bought by our teacher.

At that moment, I hated clowns and I hated the circus, whose arrival always seemed to bring stormy clouds.

Nearly forty years have passed since my first trip to the circus. As a professional journalist working for the *New York Times,* I've walked assuredly among presidents, congressional representatives, entertainers, athletes, and Americans of every race, color, creed, and political persuasion. I make enough money to go where I please and do as I choose. I've done my race proud.

And yet a little thing that happened decades ago, a painful memory that lingered long after the emotional scar healed, prevented me from experiencing one of life's simple joys.

A few years ago, when I took my six-year-old niece to

the circus, to my surprise, I found the dancing and defecating elephants rather entertaining. Beneath the painted white faces of clowns, I saw black skin as well as white. Audience members of every color and ethnicity were pulled from their seats to join in the acts. And the only balloons being handed out were those bought and paid for.

This is a book about the little things. The racial slights and indignities delivered and suffered by both sides of the black-white divide.

At a time when reports show that black Americans are doing better than ever socially, economically, and politically; when the discussion of race has entered new realms of political and scholarly debate; when an emerging body of literature is reexamining the significance or insignificance of race; and when polls and surveys show generational shifts in attitudes about race, it may be a good time to examine the simple things of life: a look at a store counter, a gesture on a crowded elevator, the translocating that takes place on a street that can cause tempers to flare, turn moods red, and unleash old emotions that sweep away rationality, wisdom, and even common humanity.

In 1997, President Clinton called for a yearlong "great and unprecedented conversation about race." "There's still some old unfinished business between blacks and whites," Mr. Clinton asserted.

But you can't legislate frank discussions. Racially sensitive exchanges seldom take place in mixed groups of ten over coffee and Danish in the executive boardroom. They happen over drinks in a corner bar, among people of the same group; at dinner tables in Harlem apartments, Upper East Side town houses, and rural trailer parks; under hair dryers in the beauty salon; or in small social settings where people's words are not likely to come back to haunt them.

Too much still remains unsaid, and too much assumed, imagined, or inferred. Blacks as well as whites are guilty of making false assumptions about the other group's motives. Resentments over seemingly small slights are felt by both. I know of blacks who refuse to relinquish an inch to whites on narrow city sidewalks, because it smacks of the bygone days of Jim Crow; of black men who feel it is the height of condescension for white men to refer to them by their first names. I remember parties that were dead on arrival the minute the words "So, what do blacks think about..." were uttered, and job offers that were rejected on the spot because a white interviewer used the word *articulate* to compliment a well-spoken black applicant. I've seen whites on a crowded subway car stand through several stops rather than approach a black youth sitting alone in a seat for two. I've seen whites called racist for acts that had little or nothing to do with race. And I've seen confusion on the faces of white Americans when blacks call each other "niggers" but dare these whites to even think about it.

This book grew out of an article I wrote for the *New York Times* in December 1997 entitled "The Little Things: Looks, stares, offhand remarks and other everyday occurrences that can ruin a black person's day." The piece also included thoughts and comments from whites who'd suffered racial slights and insults inflicted by blacks.

The story provoked a strong reaction from black and white readers. In letters to the editor, I was accused of being a racist. I was called insensitive and hypersensitive. One letter writer said I was jealous of white women. Another urged me to push my nationalist views, because he strongly believed in the separation of the races. A caller said he felt sorry for me and advised me to seek psychiatric help.

But there were also blacks and whites who related to my experiences, said they appreciated my honesty, and expressed a desire to know more. One caller who'd read the letters to the editor said he was a gay white man born and raised in Texas, now living in New York City.

"A lot of what you said was absolutely true, so don't let a few naysayers and prejudiced people dissuade you from speaking the truth," he said.

For several months I traveled the country—to Birmingham; Los Angeles; Houston; Chicago; Washington, D.C.; and Betterton, Maryland—serving as facilitator or, sometimes, mediator at focus groups that brought together black and white Americans, as well as members of other ethnic groups. Some groups were friends or coworkers; many were meeting for the first time. We gathered in friends' homes, in hotel rooms, and in my Manhattan apartment. Over drinks and hors d'oeuvres, we sat down to talk about the things that divide us as a nation. I'd hoped that in these relaxed settings, people would feel free to speak their minds without fear of retribution, misunderstanding, or racial labeling.

At times, I was shocked by the sometimes brutal honesty of my invited guests. At others, pleasantly surprised by how hard we sometimes struggle to "just get along." I spoke with blacks who voiced strong opinions about whites even though they seldom interacted with whites. I listened to whites who seemed naive in their attitudes about blacks. Many black teenagers and blacks in their twenties expressed the belief that their parents' and grandparents' generations—baby boomers and their parents—are more likely to inject race into everyday interactions. I learned quickly that mixing blacks and whites together for frank exchanges worked well enough, as long as the participants were about the same age.

Suzanne Schneider, a twenty-four-year-old black graduate student when she took part in the focus group in Betterton, Maryland, a quaint little resort town on Chesapeake Bay, may have explained why.

"Coming out of the civil rights era, they (the older generation of blacks) view our (the younger generation) current situation as one which bespoke progress (i.e. things are much better than they were before), while, on the other hand, my generation, having never experienced the hardships of the pre-civil rights era, and thus having no basis for positive comparisons, sees only that things are not good enough now—hence, our militancy and our accusations of complacency on their part."

Almost across the board there were things blacks found irritating—if not sometimes downright insulting—in their day-to-day dealings with whites. There were blacks who said that white people smelled funny, especially when they're wet; that they were arrogant and privileged; that they wouldn't know how to "wipe their behinds" were it not for blacks.

"They act like their shit don't stink," blacks have often said of whites.

At the same time, there were whites who said that blacks smelled funny, that they were hostile and angry, that they were ignorant and used slavery as an excuse for underachievement and abhorrent behavior.

Both sides felt that progress had been made between

the races, but blacks were more likely to say the progress was not enough, while whites thought significant gains had been achieved. I heard from blacks who thought integration was a mistake, and whites who saw it as America's saving grace.

Although I tried to focus on the "little things," inevitably the conversations touched on larger issues, such as affirmative action, redlining, racial quotas, slavery, and crime.

"How did the Bible get in this discussion?" H. T. Starr McCauley, a black bridal consultant, asked during a focus group in Washington, D.C.

"Do you honestly think that white folks hold these truths to be self-evident, that all men are created equal?" Evon Milton, a black woman, asked by way of answering the man's query. "The Bible is in this, sir, because whites have used the Bible to justify everything from slavery to their own superiority."

I asked the blacks present, how many, if any of them, had darkened pictures that depict Jesus Christ as Caucasian. Several hands went up, including my own.

"Now we've given him dreadlocks!" said Angelique, my twenty-one-year-old niece.

Blacks appeared more willing than whites to speak openly and honestly about their racial attitudes—perhaps because we feel we have less to lose. Whites who shared their thoughts and opinions did so, for the most part, only with the assurance of anonymity.

"We're the ones who will be labeled racist, not the blacks," said Ed, a twenty-something white male who was spending a Memorial Day weekend with his parents in Betterton. "A lot of this has nothing to do with race and everything to do with basic home training. Some people are just rude. Plain and simple."

As a child I did not know any white people. I went to all-black schools, worshiped in an all-black church, shopped in stores that catered primarily to a black clientele, and lived and played among blacks. What I knew about whites, I learned from older relatives and neighbors who worked as domestics in white peoples' houses, sleeping-car porters who supplemented their income by bowing and scraping to whites, and men like my father, who worked as a cook at the old Willard Hotel in downtown Washington. They would return home and regale us youngsters with their stories about white folks.

"You don't see them hugging and kissing on their children," my father once said. "White folks don't show that kind of outward emotion."

"They don't throw out nothing," my maternal grandmother, Ada Saunders, once said. And this from a woman who'd worked in the home of General Dwight D. Eisenhower.

Nevertheless, I wanted to be white.

Theirs was a world of privilege. Of sock hops, amusement parks, redbrick schoolhouses with new, unmarked books. Their fathers wore suits and worked in offices, and their mothers stayed at home and baked cookies. Whites were the people in the movies and on television. They were presidents, congressmen, and astronauts.

I wasn't the only little black girl who wanted to shed my blackness.

Dr. Martin Luther King Jr. once talked about how his tongue twisted and he began to stammer as he tried to explain to his six-year-old daughter why she couldn't go to the public amusement park that had just been advertised on television.

He recalls watching "tears welling up in her little eyes when she is told that Funtown is closed to colored children," and watching the "depressing clouds of inferiority begin to form in her little mental sky, [I] see her begin to distort her little personality by unconsciously developing a bitterness toward white people."

I do not remember when my awe of whites turned to fear. Had it been the tales of Emmett Till, a fourteen-year-old black boy from Chicago who was killed in Mississippi in 1955 for whistling at a white woman? I was only five when Emmett was killed. Too young to understand racism and the violence it sometimes triggered. But Emmett's death left an indelible impression on black America. Black parents told Emmett's story and recounted the subsequent acquittal by an all-white jury of the two white men accused of Emmett's murder as a kind of warning to black children to "beware of white folks."

I do know that my fear of whites turned to contempt in 1963, after four black schoolgirls were killed when a bomb exploded in their church in Birmingham, Alabama. The Ku Klux Klan claimed responsibility for the bombing. Two of the girls were my age. I no longer wanted to be white.

"Whites are evil people," the old folks in my neighborhood often said. "You can't trust them. They may act like they care about you, but they don't."

So I grew up distrusting whites. Believing all that I

had been told. I learned to keep my distance from whites. To stay with my own kind.

I'd like to think that I've changed. As a journalist I am guided by an ethic of fairness and objectivity. I live and work among whites. Some of my best friends are whites. I have stayed in their homes, shared some of my most intimate thoughts with them, and felt perfectly at ease in their presence.

Yet, I have referred to whites in derogatory terms, made fun of them, cursed them, and refused to be around them. I have played to the stereotypes about white people and sought ways to denigrate and degrade them. And I have made assumptions about whites based solely on the color of their skin.

I am not proud of my actions. But they are the result of the cumulative effect of a lifetime of racial slights and injustices suffered because of my color. Sometimes, even now, I think I'd be better-off white. God, what it must feel like to go through the day without ever once having to think about my coat of armor.

"Death by a thousand nicks," Dr. Alvin Poussaint, the noted black Harvard psychiatrist, said of the "little" things that get under our skin.

"In psychiatry, we refer to them as 'micro-aggressions,'" he said. "The things you experience every day that then add up and take their toll. Everything from being in a place and feeling invisible or ignored or questioned about your credentials or your feelings about race matters. They may seem minor, but people don't know how much this society makes blacks constantly think about being black. Many black people feel that whites don't understand them and if they bring these things up, they will be seen as oversensitive or overreacting or putting something in a racial context that is not there. These little things are things whites don't know they're doing. It's not like they're calling you a racial slur. It's being done intuitively in the way they respond to the color of your skin or don't know enough about you as a people."

This book is about perception—how blacks and whites perceive not only their own individual experiences but one another's. I consider this book a guide to understanding the racial dynamics of everyday life: the self-imposed apartheid in school cafeterias, the polarization in the workplace, the de facto segregation in housing patterns. I do not expect that everyone will agree with my opinions. Some undoubtedly will take strong exception to my points. But I have tried to be open and honest.

Doug Cregar, a white man from Chadds Ford, Pennsylvania, who participated in the Betterton focus group, explained what he hoped to find in a book such as this.

"If I'm going to Japan or India, I can get all kinds of guides about what to say and what to do. In the past few years, with more women in the workforce, there has been a lot more guidance about things that us stupid men did that affect women, and it was really good to learn about that because when you look at it from the other person's side, it's easier to understand these things and to help change your behavior. But I've not seen or heard much about relationships with blacks. I want to know, are there things I shouldn't say or things I should be aware of or sensitive to."

Mr. Cregar suggested that the answer to our racial malaise may be as simple as getting to know one another.

"If more whites knew blacks and blacks knew whites on a personal basis, that's what helps because then you learn that things are offensive in a pleasant way. The more they tease you about it, the more you learn."

This book is for the Doug Cregars of America, race notwithstanding.

1. Little Things in Public Places

I often say perception is reality, then have to spend the next few minutes explaining what I mean.

I usually do so by citing an example of extremes.

"If a police officer tells you to freeze and you reach inside your coat jacket, that cop just might shoot you. Now, you may have simply been reaching for your wallet or identification, but the cop doesn't know that. That cop might think you're reaching for a gun, and at that moment, his, or her, perception is the reality."

When it comes to race matters, perception is reality. Racial misunderstanding is mutual.

An innocent gesture can be misconstrued as a calculated insult. An entire race can be maligned or stereotyped by the inconsiderate actions of one. We form impressions and judge one another based on brief encounters or on what we believe, or are lead to believe, we know about the other group. Is the white jewelry-store owner justified in refusing admission to a black youth because he believes the young man may not be there to buy from him but to rob him? Are blacks being racially insensitive or just plain rude when they verbally lash out at whites who patronize black establishments? Does institutional racism prevent blacks from enjoying the rights and privileges routinely taken for granted by whites?

There are no easy answers to these questions, but the anecdotes provided in this chapter may give you some insight into just how much our perceptions and realities differ when it comes to the little racial things in life.

The Hair Thing

The chic woman threw back her head and ran her fingers through her long flaxen hair on a crowded elevator in Macy's department store.

"I hate it when they do that," my brother-in-law whispered. I nodded in agreement.

"If she does it again," he said, "I'm going to tell her about it!"

He sounded agitated and I gently patted his hand. The woman, still fingering her golden locks, got off on the next floor.

Later I couldn't help but wonder why such an inconsequential gesture had provoked such a strong response in both of us.

My brother-in-law, Francis Grinage, a soft-spoken, reserved black man of sixty-two, has seen the best and worst life has to offer. Unlike me, a forty-something recovering black nationalist, he has been a staunch integrationist.

Yet here we were, philosophical and political opposites, poised for a new millennium, feeling a shared sense of racial indignation over a white woman shaking her hair. What may have seemed to be a petty annoyance evoked memories of long-simmering racial slights.

We knew the woman meant no harm. She was doing what came naturally. But as blacks we understood instinctively the role hair texture has played in perceptions of beauty and privilege in America. All our lives we've been bombarded with images of white movie stars, models, and other beauty icons with long flowing hair, which has been beyond our reach. For that moment, we both saw a white woman flaunting a symbol of preference.

My brother-in-law and I are not alone in our perception. When I shared the incident with black friends of various ages, classes, and regions, they immediately understood. Kenneth Noble, a friend and *Times* colleague, is the person who originally advised me to lead with the "hair thing" in my story.

"Because it's something I think whites would never dream blacks find offensive," Kenneth explained by telephone from his Los Angeles home.

As a woman, I may have read more into the woman's gesture than my brother-in-law had. It wasn't just about hair. There's a history of suspicion, distrust, and, to a degree, envy between black and white women.

White women have been placed on pedestals well beyond black women's reach. White women have been portrayed as delicate damsels in distress, awaiting rescue by princes. Black women were mammies and sapphires—nappy-haired, big-footed, strong-willed, razor-toting heads of households, unable to please their black men.

Shanette Harris, an assistant professor of psychology at the University of Rhode Island, was once quoted as saying that black women judge themselves on the basis of hair texture and length, as well as skin tone.

White women may be hung up on their weight—you can never be too rich or too thin. But like my momma used to say: "Only a dog wants a bone."

Naah, black women obsess over hair.

"At 200 pounds, a black woman may be perfectly happy with her body but spend $100 to $300 a pop on a hair weave, denying herself both exercise and sex to protect it," Debra Dickerson wrote in an article that appeared in *Allure* magazine in September 1997. "So while white women may express self-hatred by starving themselves, for black women, self-loathing centers around what grows out of their heads."

Our mothers taught us that "a woman's hair is her shining glory."

But our kinky hair was a source of constant irritation to us and of ridicule from whites. We responded by trying to emulate white styles. We pressed, permed, geri-curled, and wove our naturally coarse hair to make it look more like the fine texture of white women's hair. Water was viewed as a natural enemy, to be avoided at all costs. Forget about blacks "not being buoyant," to quote Al Campanis, the former general manager of the L.A. Dodgers; we didn't like to swim because chlorine, water, and our coarse hair didn't mix. We ran for cover whenever it rained, fearing our permed or straightened hair would revert to a Hyde-like state. White girls, on the other hand, would jump into a pool and splash in the rain and wonder aloud why we black girls wouldn't join in the revelry.

Our hairstyles were considered pretty only if worn by white women. Bo Derek, the actress, could run through the beaches of Southern California with her blond cornrows flapping in the summer breeze and be declared a "ten." On black women's heads, cornrows became a "problem."

For that brief moment on Macy's elevator, I saw all this, and more, tossed in my face, and I responded viscerally to it.

Of the many racial slights I wrote about in my *Times* story, the hair thing seemed to strike the strongest chord among blacks and whites alike. I wondered why.

Several white women explained that it had never occurred to them that blacks would find such a thing so offensive. White women with curly hair said that white women with straight hair were more likely to be guilty of the behavior I described.

Andrew Hacker, a Queens College professor, the author of *Two Nations: Black and White, Separate, Hostile, Unequal,* and a white man who has spent decades as a sociologist studying human behavior, says blacks are more aware of the subtleties of race in America because they've been watching whites for four hundred years.

"We've already framed our perception, so we stop listening or [stop] watching behavior which might make our knowledge deeper," said Hacker. "The 'hair thing' doesn't occur to us. We don't think about it."

Indeed, whites seem to know little about black hair. For example, how soft black people's hair tends to be.

"That surprised me," said Jill Gerston Newman, one of my closest white friends. "When I touched a black friend's Afro hairstyle, I was surprised at how soft it was, because I expected it to be coarser, since that's how it appears to the naked eye."

If I had a dollar for every time I'd been asked by a white person whether or not I'd cut my hair when all I'd done was wash it, I'd have enough to buy shampoo to last a lifetime. Here's what's up with that:

When a black woman with nappy hair, like mine, washes her hair, it often looks as though it's been cut, because it's usually tighter and closer to the scalp. After a few days, the hair loosens, becoming fuller, with more natural body.

A little thing Ronald Prince, my hairstylist, has noticed is that while most black beauticians can do white hair, few white stylists know how to style black hair.

"We learned how to do all types of hair in cosmetology school," explained Mr. Prince. "White hair was the standard we were taught to emulate, whether it was using a straightening comb or through chemical relaxers. It's been that way since the early 1900s, when Madam C. J. Walker started training black beauticians. Until recently many of the white cosmetology schools did not train students about the art of styling black people's hair, even though it's common knowledge that black women spend more money getting their hair styled than white women, in part because black hair requires more care."

There were blacks, however—especially black women—who said they didn't "get" the hair thing.

"Why would something like that bother you?" a friend asked. "I couldn't care less about white women and their hair."

"Then why do you perm your hair, Mom?" her twenty-something son asked. "If you don't care, why don't you wear your hair in a natural?"

"Because I like my hair permed, and no, I don't think that's emulating white women, because they perm their hair, too!"

Several black men said the hair thing also rubbed them the wrong way, whether it was a female or a male doing the hair shaking.

"Yeah, white men do that, too," said Deriek, my twenty-five-year-old nephew. "They wear their hair down to here"—he pointed his finger below the nape of his neck—"and get on crowded subways or buses and start running their fingers through their hair within a hair of your face."

A white woman who read my story telephoned to say such behavior goes beyond race.

"It's downright rude," she suggested. "Hair is dirty, and you shouldn't go around shaking your hair in people's faces."

Now, I do not want anyone to think that I'm so sensitive that any white woman who stands near me and starts shaking her hair must bear the brunt of a nation's troubled racial past. It's not what's done; it's how you do it.

I have no problem with whites who simply brush their hair behind their ears or away from their faces. Everyone knows that hair can be annoying to the skin and an irritant to eyes. A simple strand can ruin a good meal, and God forbid it gets in your mouth!

But this flipping and flopping and slinging is just a tad too cute for comfort! . . .

Hair has long played a significant role in shaping racial attitudes and perceptions. Since I was a child I have been hearing blacks say that white people's hair, when wet, smells the same as wet dogs. That was the 1950s. I'm still hearing it today from smart, intelligent, sophisticated black Americans, some who've dated whites.

In fact, when I asked people to fill out a questionnaire and list three things they've heard said about white people and vice versa, half of the blacks listed the "wet-hair thing."

David K. Shipler, former *Times* colleague, friend, and author of *A Country of Strangers: Blacks and Whites in America,* believes that blacks are aware of their own prejudices and they more or less told him as much during his interviews with them while researching his book.

"This whole thing that you shouldn't get near a white person who's been out in the rain because their hair stinks was first told to me by a black woman in Washington, D.C. Then a black columnist at the Detroit *Free Press* said she'd heard this," Shipler told me.

A colleague of mine who's from Canada said he'd been listening to a program on WCBS radio, in the spring of 1998, on blacks' and whites' perceptions of each other.

"I was shocked," he told me. "Do you know what blacks said?"

He didn't wait for my answer.

"They said that white people smell funny."

I offered a smile and a reply. "Now what exactly did they say you smell like?"

"Well, they said when we get wet, we smell like wet dogs. You knew, didn't you?"

I asked if he knew what white people said about how blacks smell. "Well," I answered myself, "I've been told that we blacks smell like dirt, ammonia, sweat—and our hair smells like sour milk."

"Maybe what people are smelling is dirty hair?" he suggested.

He probably has a point.

I, for one, can personally attest to the fact that not all white people's hair smells funny when wet. Jill has curly shoulder-length hair. During her many stays at my Manhattan apartment, she has often washed her hair. On one such occasion I was applying makeup in the bathroom standing behind Jill, while she was blow-drying her wet hair. Did I smell anything funny? You bet I did! My $25-a-bottle Frédéric Fekkai shampoo and the $25-a-bottle conditioner.

The Invisible Black Man or Woman

A black person has been standing in line in, say, a restaurant—or in a movie theater or at a department store's cosmetic counter. A white person arrives and cuts the line with not so much as an "excuse me," completely ignoring the black person and asking the maître d'hôtel how long the wait will be for a table—then responds to the black's objections by "innocently" saying something like, "Oh, were you waiting for a table?" Worse, is when they say, "Oh, I didn't see you standing there!"

Whites might view these incidents simply as people behaving badly. But many blacks see them as the typical actions and attitudes of a people that still believe, as the

Supreme Court declared in its 1857 Dred Scott ruling, "blacks were so inferior they had no rights that a white man was bound to respect."

It may be history, the past, and a different time and place to whites, but for many blacks—some too young to remember and others too old to forget—changes are slow to develop.

"Blacks tend to see history as currently relevant and echoing in the present," said David Shipler. Mr. Shipler, who is white, said, "Whites tend to see history as the past and therefore irrelevant and not governing the present at all. That is one of the major fissures between blacks and whites, and it results in lots of misunderstandings."

"They don't see us," Dennis Lynch, my nephew-in-law, said during an interview in the summer of 1998, some twenty-five years after Dr. King referred to the invisibility black Americans live with as a "sense of nobodiness." Mr. Lynch said, "Whites still look at us as though we are children of a lesser God. No need to respect our rights or wishes. I've seen them acknowledge people's dogs before they acknowledge us."

Dr. Ted Manley, a professor of sociology at DePaul University in Chicago, said another manifestation of the invisibility blacks experience is what he calls "the white eye-treatment."

"White people do not maintain eye contact with black people, even if they know you. So they give you this white eye-treatment," said Dr. Manley, whose broad, six-foot-three dreadlocked presence is hard to overlook. Nevertheless, Dr. Manley recalled close encounters with white colleagues that left him feeling otherworldly.

"I live in Oak Park, Illinois, and I see one of my female

colleagues, a white woman, with her husband, walking down the street. I looked right at her and she kept walking. The next time I saw her on campus, I said, 'I saw you and I know you saw me, and you gave me the white eye-treatment.' And she looked at me and said she was very sorry. She said she was caught up in the conversation with her husband and didn't have the chance to acknowledge me."

Dr. Manley said he accepted her explanation because he felt it was her way of apologizing and offering an excuse for her behavior. Did he believe her?

"No," he said. He explained why by citing another white eye-treatment incident.

"Just the other day, a white male colleague, who teaches at DePaul and works in the same building with me, got on the train with me at the same stop, looked me in the eye, and said nothing. I said to myself, *I'm not going to say hi,* because too many times I have to be the one to initiate. So I just looked at him and turned away."

The two rode together for twenty minutes, got off the train, stood side by side on an escalator, and walked across campus within steps of one another. Only when Dr. Manley held open the door to the library for his white colleague did the man say hello.

"He didn't acknowledge me on the train platform, on the train, on the escalator, or walking across campus. I'm carrying a briefcase and all the academic paraphernalia he had, but he didn't acknowledge me."

To speak or not to speak, that is the question. We—blacks, that is—have all felt the sting of being ignored in public by whites we work with or attend school with. Some of us have developed a sixth sense about such matters. We'd rather cross the street or avert eye contact than be dissed in public by whites we consider colleagues, classmates, even friends.

Racism, many blacks believe, has rendered us invisible. We are a people without a name or a country; a people lacking an identity, other than that given us by whites; faceless entities placed on earth to serve the white man. In her autobiography, *Passion and Prejudice,* Sallie Bingham, the newspaper heiress wrote: "Blacks, I realized, were simply invisible to most white people, except as a pair of hands offering a drink on a silver tray."

The black writer and educator bell hooks offered a contemporary point of view in her essay "Representations of Whiteness in the Black Imagination."

> Since most white people do not have to "see" black people (constantly appearing on billboards, television, movies, in magazines, etc.) and they do not need to be ever on guard nor to observe black people to be safe, they can live as though black people are invisible, and they can imagine that they are also invisible to blacks.

Most whites, for example, do not understand why blacks become offended when white people say they don't see color.

"Well, I don't see color," a white colleague said once during a diversity meeting.

"Then you must be color-blind," several blacks retorted.

My colleague tried earnestly to defend his point of view. "I see people as people," he explained.

"But don't you see how offensive that sounds when you take away a vital part of a person's identity?" he was asked.

"No," he said.

Removing color from the race equation may seem like a noble gesture. It is not.

In an essay examining the difference in dialogues between black and white women, Marsha Houston, a professor at Tulane University, said such comments "suggest that the speaker regards blackness as something negative, a problem that one 'can't help' and, therefore, as something that one's white friends should overlook. It denies the possibility that blackness could be something to be valued, even celebrated."

Take the same observation and replace the word *color* with *gender* or *age* and maybe you'll get the point. How unfortunate women might feel if men said they "don't see gender." How harmful it would be to elderly people if we said we "don't see age."

Dr. Manley says he carries a box of crayons in his briefcase as a way to illustrate the absurdity of such observations.

"Whenever people say that to me, I throw the crayons out there and say: 'Do you see those?' "

Once, he recalled, he did that to a white colleague who, indeed, did not see the difference. "I'm color-blind," the colleague told Dr. Manley.

"He truly couldn't distinquish yellow or black or red. Now that's legitimate. His was physical, but it's a rarity. It's better for you to say, 'color doesn't matter' than 'I don't see color.' 'Color doesn't matter' tells me that you know I'm a person of color but that's not going to affect how you're going to look at me."

Those who say such things may mean no harm. Maybe they are trying to show racial sensitivity. I believe there are whites who look beyond race and see blacks as individuals. Then why not say that? Why not say that you, in the words of Dr. King, "judge people by the content of their character, not the color of their skin."

Because that's easier said than done.

Every time a white woman clutches her purse when a black man sits next to her on a subway car or a bus; every time a white salesperson chooses to follow a black shopper who enters a store, instead of a white customer; every time a police officer pulls over a black motorist just because he's driving a luxury car, whites are making judgments based on race rather than character.

"I'm walking down the street in Glendale, California, with a sundress on, a plain white baseball cap, and nothing more in my hand than my bank card," said Roslyn Myles, a young black woman who lives in Los Angeles. "I see this white woman up the street heading in my direction suddenly cross the street and continue walking in the direction from which I was heading. I realized then that she must have seen me coming and decided to cross the street. I was thinking, *Did you look at me? If I were blond, wearing a plain white baseball cap, and in a sundress, would you cross over? It's not like I could be hiding a weapon anywhere, so why are you afraid of me?*

A black computer-operator in Washington shared the following story to illustrate the daily frustrations blacks often encounter in public places.

"I was standing outside the Four Seasons Hotel in Washington one day, dressed in a pinstriped suit, wearing wing-tip shoes, and holding a briefcase in my hand," he said. "An older white woman hops out of a taxicab and tells me that her suitcases are in the trunk of the cab. When I didn't budge, she stared at me and pointed toward the back of the taxi and said loudly, 'My suitcases are in the trunk; take them to the front desk.' "

The man proceeded to remove the suitcases, place them in front of the hotel, and hop into the cab. From the backseat of the cab, he could see the woman standing there shaking her fist as the taxi pulled off.

Blacks believe whites do not want to see them because in our faces whites see America's ugly past; they see a people wronged by their doing; they see America's only redeeming grace—a lesser people may have succumbed. Whites can't bear to look us in the eye without feeling guilty.

Every day, the present and past collide. Sometimes bottled-up despair and anger uncork bad behavior by blacks.

One such incident occurred on a crowded Manhattan sidewalk when a well-dressed middle-aged black man was headed on a collision course with a group of young whites so engrossed in conversation and hand-holding that they failed—or didn't bother—to see him heading in their direction. When the group and the man finally reached a point of no return, the man pushed his way through with such force, he nearly knocked one of the women into the gutter. The group looked dumbfounded and probably dismissed the close encounter as just another rude only–in–New York occurrence.

Several of my black friends said that if the brother had

shoved *them* out of the way, psychological baggage or not, they would have been in his face.

I, however, saw a black man striking a blow for his ancestors. For all the times they were forced into oncoming traffic, threatened with beatings, or subjected to tongue-lashings for failing to step aside with haste for white folks. I saw a brother fed up with eating crow, as in *Jim*.

In the days of Jim Crow, blacks were supposed to "stay in their place," which was at the back of the bus, at the back door, and clear of whites on sidewalks. This legal system of segregation prohibited racial intermarriage, required separate seating in trains, buses, theaters, libraries, and stores, as well as separate schools, restrooms, public drinking fountains, parks, and other public facilities.

The name *Jim Crow* was taken from a blackface singing-dancing comedy characterization that portrayed black males as childlike, lazy, irresponsible, pleasure seeking, and happy—a stereotype passed down through the ages. Although abolished in the mid 1960s, the laws' painful legacy continues to live in our collective memory, dictating behavior—good and bad—on our part.

I know of blacks—male and female, young and old—who refuse to give up their paid-for seat in the front of the bus for any white person, regardless of age, gender, or disability.

"Hey, let one of *them* get up for 'em," was how Tyrone, a black teenager from Brooklyn, explained his defiant apolitical stand.

Would he relinquish his seat on the bus for an elderly black woman?

"Yeah, 'cause I know she'd probably had to put up

with a lot more shit from the white man than I ever had. And if I didn't and my grandmother found out, she smack me upside my head!"

"Black-white interaction today is being renegotiated," said Dr. Joe R. Feagin, a white professor of sociology at the University of Florida, who has written extensively on racial and ethnic relations, including *Living with Racism: the Black Middle-Class Experience,* a book he coauthored with Melvin P. Sikes. "Prior to societal desegregation, blacks typically expected to respond to discriminating whites with great deference, stepping off sidewalks when whites went by. Since the 1960s, whites have encountered a majority of blacks who do not respond with deference when faced with discrimination. They may withdraw or deny, but they do not shuffle."

Years of navigating racial minefields has given blacks what Dr. Feagin calls "the second eye," which one black described as "the ability to sense prejudice or discrimination even in a tone of voice, a look, or a gesture."

For example, my niece, Antoinette Williams Lynch, who is thirty, spent several hours a day commuting from her apartment in Silver Spring to her former job in Lynchburg, Virginia. She traveled along the Beltway, a heavily trafficked superhighway ringing the Washington metropolitan area. She kept her "second eye" on the lookout for "two or more T-shirted white boys with long stringy hair, riding in a pickup truck."

"They can be the worst nightmare for a black woman traveling the roads alone at night," she explained, adding that her fear was based on reality as well as on popular cultural perceptions. "You just know they got a case of

Rolling Rock bottleneck beers on the floor and are ready to have some fun... at your expense."

Quinn Walker of Washington said his second eye told him it was time to leave a white fraternity party thrown at the predominantly white college he attended, after the white kids began drinking too much, trashing the frat house, and encouraging the girls present to strip.

"All I needed was for the cops to show up and see my black face," said Quinn. "I would have been in jail, and the whites kids would have been sent to their rooms."

Although some of the whites at the focus group in Washington took issue with his opinion, Quinn said he'd relied on his "gut feelings," the ones black men often get when white boys start fooling around.

"We can't afford to have anything go wrong, because we're going to be blamed," explained Quinn.

Black Americans are forever fighting that "degenerating sense of nobodiness." We do not expect whites to understand what that must feel like. There are whites, however, who do.

Francine Cregar, a white mother of three sons, who participated in the Betterton focus group, recalled the numerous times she had been relegated to the realm of invisibility when she told people she was a housewife.

"I think that for fifteen minutes I wasn't important, but for a black person maybe it's twenty-four hours that they weren't important, only because they are black," she said. "So when you talk about a black person feeling he or she may be ignored, I can understand a little bit more. I say I'm a Lucite person. People look at me and then they look past me because on the surface I'm a mother and the wife of someone."

The Look

It starts with an expressionless stare. The eyes begin to squint. The mouth opens slightly. And you know you're being analyzed or sized up. It is a look whites often give blacks who don't fit the composite notion of a Negro. If you're black and you've lived long enough, you know "the look" when you see it. Often it comes right after you say what you do professionally.

"Really! So how long have you worked for..."

Sometimes the look appears when blacks say they have attended a college or university that is not a historically black school.

"You went to *Harvard*?"

"I've been at parties and had white people ask me what high school I attended," Dr. Poussaint of Harvard recalled. "It's disconcerting, especially at my age, because you know they're asking to find out your history and how you arrived at this point."

A young black female television executive said she gets the look whenever she flies first class. Which is often, since her job pays for first-class passage during her numerous cross-country business trips.

"Especially if I don't preboard," said the woman, who agreed to talk over lunch at a midtown Manhattan restaurant, provided I not divulge her name or business affiliation. "Without so much as glancing at my ticket, the flight attendants usually direct me to the right, toward the back of the plane. When I say, 'I think seat 2A is to the front,' they get all red-faced, then ask to see my ticket. As if I'd lie about something like that."

Ronald Frazier, a black middle-class Washington executive with two baby Benzes in the garage, often gets it when he offers a business idea or strategy that will save companies that he deals with time or money.

"You can see the look on their faces," said Mr. Frazier, referring to the whites in the room. "You can tell they wished they had come up with the idea instead of me. They're trying to figure out how this nigger came up with such a brilliant idea."

I've jokingly told black friends that *the look* means "you don't compute"—in other words, the sum of your life doesn't quite add up to the parts: black, American, minority member, second-class citizen. And now, you do *what?* Blackness as an aberration.

For years, Cynthia Jones says, she got the look whenever she told whites she was reared on the Upper West Side of Manhattan, attended private schools, graduated from Oberlin College, and received her law degree from the Benjamin N. Cardozo School of Law at Yeshiva University.

"You went to *Oberlin*?" Ms. Jones said, marking the tone and incredulity of some of her white inquisitors. "Or they'd say to my mother, '*Your* daughter goes there?' "

Roslyn Myles, who works as a commercial decorator on movie sets, said she once got the look from a young white receptionist at a major motion picture studio in Los Angeles. Ms. Myles had arrived, briefcase in one hand, camera in the other, for an assignment.

"She looks at me and says, 'Talent is down the hall,' " Ms. Myles recalled. "I said, 'No, I'm here to see Mr. Such-and-Such.' And she says, 'No, honey, talent is down the hall.' Then I raise my voice and say, 'I'm the decorator.' And the receptionist says, 'Oh, you want to see Mr. Such-and-Such.' Well, that's what I said when I came in."

So her boss wouldn't make the same mistake, the receptionist introduced Ms. Myles by giving her occupation first, then by her name: "Mr. Such-and-Such, this is the decorator, Roslyn Myles."

"It doesn't bug me so much that they assume I'm talent, as if that's what I'm here for and no other purpose," Ms. Myles added. "But when you're carrying a briefcase and strutting a camera? When does someone show up like that for talent?"

Denzel Washington, one of Hollywood's bona fide superstars who happens to be black, told Edward Bradley, the *60 Minutes* correspondent, that he got a look from a little white lady in a public elevator. It was the kind of please-don't-harm-me look black men often get from little old white ladies.

"I could crush her with my wallet and she's—," said Washington, recoiling from the cameras to show how the woman responded to his presence. Washington, who gets about $15 million per film, was laughing as he recalled the incident.

I've gotten the look for, what, twenty-five years, now. As long as I've worked for the *Times*, I've seen that expressionless daze come across the faces of white public-relations types when I walk into a press conference and they look up and see my brown face.

"Are you here for the press conference?"

And I'm thinking... *No, I always get dressed to the nines and all made up, then just show up, uninvited, to corporate boardrooms to listen to dignitaries talk about life, liberty, the pursuit of happiness.*

Instead I answer, "Yes, I'm Lena Williams."

That usually brings a long puzzled look, then "Lena Williams and, and . . . who are you with, Ms. Williams?" *We didn't invite you here because you're the seventh child of Ralph and Lena Williams, honey. Tell us what media organization you're affiliated with or drop the Danish.*

"I'm with the *Times,*" I hear myself saying.

"Oh, *Ms. Williams*!" A smile comes across the glazed eyes and I'm given that quick once-over, from head to toe, then graciously ushered in to meet the chief executive officer.

For that moment I know what white privilege feels like, and believe me, it feels good! I know the public-relations officer is only doing his or her job.

And I must say that I'm not given the look nearly as often as in years past. But it still happens. I still get interrogated by whites when I tell them I work for the *Times.*

"So, how long have you worked there?"

"Did you go to journalism school for that?" (What's up with that?)

Whenever blacks talk about their travails in public, whites often respond by recounting their own trials and tribulations in the workaday world.

"That happened to me once."

Whites who do so may be guilty of nothing more than showing empathy for our suffering daily trespasses. Blacks, however, often view such remarks as an attempt to undermine or negate racism as a possible factor for such behavior.

Trust me, America, the slights being discussed here are not just once or twice a year. We're looking at lifetimes of the slings and arrows of outrageous fortune.

"It's so hard to convince them," said Dr. Manley. "And

why should we have to? That brings out the reason why there is the need for blacks and whites to talk to each other. Because if we don't, we're heading on a crash course."

G. P. of Los Angeles, one of several blacks who requested anonymity because they did not want to alienate white friends or colleagues, listened patiently during a focus group in Los Angeles while other blacks around her recalled racially insensitive encounters with whites that had left them smarting to that day. Then G. P., whose job required her to interact daily with members of other races and so had learned to navigate the racial and ethnic minefield with an incredible adeptness, told of an incident that defied logic, even though she tried to find a reason for it.

"I bank at Bank of America. I have my checking account there, my savings account, my bankcard with them, my money market with them; and they hold the mortgage to my house. Whatever branch office I go into, unless it's the branch in the same offices where I work, I'm asked to show a driver's license, and I must also punch in my ATM card PIN in order to cash a check. In August I went to a BofA because I thought the bank had made a mistake and double credited a deposit I'd made at a different branch. When I arrived I was told that I had a different balance and that several of my checks had bounced."

What G. P. later learned was that someone in the San Francisco Bay Area, who had neither her checking nor savings deposit slips, obtained her account number and was able to deposit $3,500 in fraudulent checks into her account and then get cash back on the transactions.

So what was G. P.'s point?

"My point was, I had banked with BofA for seventeen

years, and yet anywhere I went to conduct bank business I am challenged, as a black woman, but someone was able to get money from my account with no questions asked."

When asked what she would say to those whites who may have had similar experiences, G. P. replied, "I'm always asked for identification. This isn't once or twice."

Regardless of gender, socioeconomic class, status, or style, no black American is ever totally immune from knee-jerk racism, be they sanitation worker or U.S. representative.

Congresswoman Cynthia McKinney, a black Democrat from Georgia, complained of receiving "disparate treatment" at the White House on two different occasions, one in December 1996, the second in May 1998.

Describing the May incident, Ms. McKinney said security guards and Secret Service agents failed to recognize or acknowledge that she was a member of Congress. She said that when she arrived at the northwest gate of the White House with her white assistant, a white guard refused to address his questions to her and assumed that her twenty-three-year-old staff assistant was the member of Congress.

"I am absolutely sick and tired of having to have my appearance at the White House validated by white people," Ms. McKinney wrote in a letter to President Clinton. "Unfortunately, disparate treatment of non-Caucasians has been a part of the 'history' with the White House for the past six years I have been in office," she wrote.

According to Ms. McKinney, once she got to the West Wing, she was stopped again by a security guard and was allowed to move on only after Representative James P. Moran, a white Virginia Democrat, vouched for her.

Can You Afford to Buy This?

Just once, I'd love to go shopping dressed down.

I never do it. Can't afford to. For me, and millions of other blacks, going shopping in the white world requires a certain pedigree. We must be dressed properly—preferably in business attire—and have our wallets filled with money or credit cards, lest we be mistaken for a shoplifter.

Even looking our best doesn't necessarily mean that we will not be slighted, falsely accused, or followed around the store like a common thief. No matter how affluent and influential a black person may be, he or she cannot escape the stigma of being black, especially when shopping.

Oprah Winfrey, the popular talk show host, was once denied access to a boutique on Madison Avenue. Debbie Allen, the actress, director, and choreographer, was refused service at a boutique on Rodeo Drive because the saleswoman thought Ms. Allen could not afford to purchase merchandise in the store. Ms. Allen bought the item to prove a point, only to return it the next day after she realized the offending salesclerk received a substantial commission from the purchase.

Celeste Goring-Johnson, a forty-two-year-old mother of three, from Brooklyn, wasn't given the chance to show indignation like Ms. Allen's. In September 1999, Mrs. Goring-Johnson, who is black, was arrested, taken away in handcuffs, and strip-searched because a Brooklyn jewelry-store owner suspected she had stolen a diamond ring.

Mrs. Goring-Johnson denied taking the ring. As it turned out, the only evidence against her was that she was the only customer near the counter when the jewelry-store owner noticed the ring was missing.

After Mrs. Goring-Johnson was released from custody, she contacted a lawyer and submitted to a lie detector test, which she passed. The district attorney's office did not prosecute the case.

"Blacks are seen as shoplifters, as unclean, as disreputable poor," Dr. Feagin said to me when I was researching a story on the bias blacks experience in public places. "No matter how affluent and influential, a black person cannot escape the stigma of being black even while relaxing or shopping."

Having occasionally experienced poor service themselves, many whites accuse blacks of being paranoid in seeing racism in such incidents, noted Dr. Feagin. "One more aspect of the burden of being black is having to defend one's understanding of events to white acquaintances without being labeled a racially paranoid," he wrote in his 1994 book.

Besides, blacks have never received good service in businesses operated by whites. Not even during the days when "customers were always right."

When venturing into white worlds, most blacks take certain precautions.

I always allow an extra fifteen minutes to hail a cab to get to and from the airport or to assignments, when possible. I do so knowing full well that I am likely to be passed by for a white fare. In order to reach my destination in time, I must allow adequate time to account for other people's ignorance.

The "cab thing" has become an accepted fact of black life in America, especially in New York City, where comedians have turned the humiliating experience into a gag.

But Danny Glover, the black actor, widely recognized for his starring role in the *Lethal Weapon* movie series, wasn't laughing on October 9, 1999, when several taxis refused to pick him up in lower Manhattan. Glover, who stands six feet four inches, was forced to wait in the shadows while his daughter hailed a cabbie and then talked the driver into opening the door. Adding insult to injury, Glover, who has a bad hip, said the driver refused his request to ride in the front seat, a request city officials said should have been honored.

Glover did what so many other blacks before him had done: filed a discrimination complaint. But unlike us ordinary black folks, Glover was a celebrity. When he talked, the media listened. Less than a week after Glover filed his complaint on Nov. 3, New York City mayor Rudolph W. Giuliani announced an undercover operation in which minority-member police officers would pose as taxi patrons and issue tickets to cabbies who refused to pick them up. Rudy—as we New Yorkers like to call him—even threatened to confiscate the cabs of violators.

In a two-week period less than two dozen summonses were issued, leading to speculation among some whites that blacks had exaggerated the depth of the problem. I think not. I, for one, could attest to the fact that shortly after the undercover sting was announced, on two separate occasions cabdrivers—one white, another Pakistani—actually passed up two white fares, one a man, the other a woman, to pick me up. I wondered if the cabbies thought I was an undercover cop posing as a reporter with a *New York Times* briefcase and brown trench coat. One cabbie knew he'd blown my cover for sure when he dropped me off in front of my Manhattan apartment building, which is located across from the twentieth police precinct.

"Oh, so you live around here, huh?" he said with a smug grin on his face.

"Why?" I shot back. "You don't get many black passengers going to the Upper West Side? I guess we all live in Harlem, huh?" I told him to keep the change, then watched him through the glass of my front door, watching me enter my apartment.

As soon as it appeared that blacks' feelings had been assuaged by the mayor's public-relations gesture, the sting was lifted and cabdrivers returned to their old pass-up-blacks ways.

But selective discrimination in public isn't just a cabby thing.

I choose restaurants not by the number of stars they receive from critics but by what other blacks have experienced dining at these establishments. If blacks say they have been treated with respect and given proper service, and were not stared at by other patrons and not forced to sit near the kitchen, I don't care whether the food is mediocre or downright bad, I'll go. I always tip above the recommended 15 percent, so that the next black patron will not be abused because of my oversight.

Another little thing I often do to insure proper service is use my corporate American Express card, rather than my own personal credit card. My corporate card bears my name and that of my employer on it. It carries more clout than my Citibank Visa card, which bears only my name. It makes whites look beyond the black face at the table and see, instead, a status symbol, one respected by whites.

The luxury of not having to worry about such superficial matters was one I was not afforded, because of my skin. Every now and then I imagine how wonderful it must feel to be able to go out in public looking any way one chooses and be treated and accepted as a first-class citizen. The color of my money should be the only color that matters when it comes to such matters.

But long before we can "show them the money," we have to show them our faces. And Kermit the Frog thinks it's not easy being green. Try being black.

You Won't Find Us up in Those Woods

I like rustic settings. Out-of-the-way places, far from madding crowds. Nature. The woods. My idea of a real retreat.

Some of my relatives and black friends think I'm nuts.

"Girl, you don't mind going up in those woods alone?" my sister Ada asked about a summerhouse in Old Chatham, New York, I rented with two of my black girlfriends, Diane Camper and Lydia Pitts.

Ms. Camper, a former *Times* journalist, and Ms. Pitts, a corporate lawyer, wanted to find a refuge from the hustle and bustle of New York. We could have done the "black thang" and rented a place in Sag Harbor, the popular black summer resort community on Long Island, but we dared to be different. Besides, in Sag Harbor we were bound to see some of the same folks we worked with.

But neither logic nor reasonable explanations would assuage my sister's concerns.

"How many black folks do you run into up there? Huh? How many? You and your girlfriends, that's all.

They'll get you guys back up in those woods one day and we'll never find the remains."

Surely, she jests!

Not really. For most Americans, planning a vacation involves a few basic considerations: where to go, where to stay, transportation, and costs. But for many black Americans there is often another factor to be considered: Will they be the only blacks there?

While most blacks no longer feel uncomfortable about moving in predominantly white circles at work or in their communities, the thought of being on vacation alone, among strangers, in an unfamiliar city, town, or resort, even for a short period, can make travel less exciting, even disquieting.

Many blacks choose the places where they vacation more carefully than some people choose their schools or homes. The idea that some white bogeyman lurks in the woods is older than the Ku Klux Klan itself. There is safety in numbers, we have been told. And there is nothing more vulnerable than an isolated black man or woman.

Bob Hayes wrote of the phenomenon in his book *The Black American Travel Guide,* published in 1971 by Straight Arrow Books. "This Guide alleviates destination facility hazards: it minimizes the problems of deciding where to go and of worrying about what's going to happen when you get there."

Whites may find such attitudes silly, to say the least. Although Diane, Lydia, and I never experienced one racist encounter or incident in the four years we ventured into the woods in Old Chatham, tales abound of blacks who ventured into predominantly white or all-white areas and found no room at the inn, no place at the table, and unwelcoming stares.

The author bell hooks talks about the fear many blacks experience traveling the world. "To travel, I must always move through fear, confront terror," she wrote in an essay. "It helps to be able to link this individual experience to the collective journeying of black people, to the Middle Passage, to the mass migration of southern black folks to northern cities in the early part of the 20th century."

Ms. hooks recalls traveling to Italy, as an invited guest of a member of Parliament to give a speech on racism and feminism, and being interrogated for hours by white officials. She remembers being strip-searched by French officials, who were stopping black people to make sure they were not illegal immigrants.

For years there weren't many blacks seen on ski slopes or hiking in the mountains. We didn't rent secluded little cottages in Maine or Vermont. We didn't run off to the Grand Canyon or Mount Rushmore. We stuck to safe havens, places where black folks hung out: Sag Harbor, Oak Bluffs, Maryland's Eastern Shore, the Caribbean, cities with sizable black populations. Vacations are supposed to be a time of fun and relaxation. Who needs grief? All too often blacks who tried to do the "white thing" by vacationing in places whites frequented were sorely disappointed.

It can also happen in the least likely places. Disney World, for example.

When I went with some members of my family to Disney World in Orlando in 1980, I was less than pleased with the reception we received in some of the restaurants and at the hotel. We were a party of nine—large by some standards—but we'd heard of busloads of tourists who traveled to Orlando each year and managed to be seated together, or near one another, while dining. These tourists didn't return to the hotel after four, five, or six hours at the park only to find their rooms still unserviced. But the Williamses found exactly that.

After being seated for breakfast near the kitchen door two days running, we got loud (ethnicity has its advantages) and demanded a table by the window. We went to the front desk before heading out to the park and asked politely that our rooms be made up by the time we returned, noting in increasingly louder volume that that had not been the case the previous two days. Yes, it drew stares from whites in our immediate vicinity, but we stared back and they got the message. In other words we played to the stereotype about us blacks folks being the violent type.

Although I no longer make my vacation plans based on the black quotient of vacationers, many of my black friends and relatives continue to do so. But I know I must prepare myself for that old familiar feeling of being "a fly in a pan of milk." It doesn't matter that I may be treated kindly and greeted with open arms in these places. The sense of loneliness and isolation can be so overbearing that I often have second thoughts.

In the spring of 1998 I went to a spa in Glenwood Springs, Colorado, for three days. One morning I had breakfast at the Hotel Colorado, a historic landmark. I was seated alone at a table that morning, reading a newspaper. Everything was fine, including the food and the service.

Later that evening, I was in a drugstore, and a white couple standing in line behind me kept looking at me. Finally the man touched me on the shoulder. "Weren't you having breakfast this morning at the Hotel Colorado?" he asked.

I nodded yes.

"We thought we saw you."

I guess they did, since I was the only black face I'd seen around town, except for a masseur at the spa. I didn't take offense. The couple, who were from Denver, and I had an interesting chat. Later when I shared the story with my black friends, though, I tried to explain how weird it was. Think for a minute. What are the chances of a stranger recognizing another stranger he or she had seen briefly from afar in a town, the size of, say Alexandria, Virginia? Especially when the second encounter took place hours later in a completely different location. Then again, when you're the only black face in town, anonymity isn't an option.

And another little thing. There's nothing worse than enjoying yourself vacationing overseas sans the race factor only to return home to America and be confronted with racial stereotyping from white customs officials.

I know what whites are going to say—that it happens to them, too. Or that blacks are stopped and questioned when entering the country because we fit the profile of drug smugglers.

Really? My girlfriend Isabel Wilkerson, who in 1994 as a national correspondent for the *Times* became the first black journalist to win a Pulitzer Prize for reporting, is a petite woman who seldom ventures anywhere, even to the corner grocery store for milk, without being fashionably attired. She wears her shoulder-length hair in a conservative style, speaks softly, and conducts herself with the utmost discretion. Doesn't sound like a drug runner to me. But in August 1996 she and her husband, Dr. Roderick Watts, a professor of psychology at DePaul University, were stopped by customs officials upon their arrival at Chicago's O'Hare Airport following a vacation in Italy. For forty-five minutes the couple was detained, interrogated, and forced to unlock bags that had gone unopened during their brief stopover at Heathrow Airport in London.

"What was so upsetting was that the day we left for Italy was the same day TWA flight 800 had fallen in flames into the Atlantic Ocean shortly after takeoff from [Kennedy] Airport in New York," said Ms. Wilkerson. "If you recall, they suspected the plane was downed by terrorist attack, possibly a ground-to-air missile. While we were still in Italy, a bomb exploded in Atlanta during the Olympic Games. I wasn't angry because we were stopped and searched. But [because] we were the only passengers on our flight who were detained."

Ms. Wilkerson wrote about the experience in an op-ed piece for the *New York Times*. Days after the incident, she was told by a senior customs official that the process should have taken all of thirty seconds. "Once you establish your U.S. citizenry, that's it, you're done," the official told Ms. Wilkerson.

Imagine what that must feel like. To travel the world and be treated like a first-class citizen because of your American passport, only to return home and be manhandled like some illegal stowaway. Ain't this our country, too? Sometimes we are left to wonder.

"Whenever we start to complain about our treatment in America, some racist white will say: 'Go back to Africa,'" said Mrs. Mitchell of Washington. "We'll go back to Africa when they go back to Europe or the Eastern Bloc nations where they emigrated from. All these whites suffering from selective amnesia seem to have forgotten that Native Americans were here when Christopher Columbus 'discovered' America. It was never lost.

"Blacks helped build this country. We helped make it the great nation it is today. We'll leave when white folks leave, and maybe not even then."

I've always been ambivalent about my nationality. When I was young I loved being an American. I was as American as Mom, apple pie, and Chevrolet. In school we pledged allegiance to the flag, staged plays on the American Revolution, with little black girls playing the role of Betsy Ross and little black boys reciting the ride of Paul Revere. I believed in democracy, saw communism as a threat to the world, and thought Africa was synonymous with the jungle.

Then black power got hold of me. I danced on the flag and grew to despise the hypocrisy of a nation that preached freedom and justice for all while oppressing a segment of its population. A nation that wanted to dictate policy to communist regimes in the former Soviet Union and Cuba but refused to right its own wrongs. Black soldiers fought and died in wars in Germany, Japan, and Vietnam, only to return home to be called niggers. Forget integration; I believed in a separate black state.

When did I change back? When did I begin to hear myself say how disappointed I was that children today do not know the pledge of allegiance? When did I begin to routinely refer to myself as a black American?

Was it 1978, when I and two friends went to Paris and heard Parisians instinctively refer to us as Americans. Not black. Not Negroes. Americans. Or was it in 1983 while visiting London, where British citizens remarked on my American accent.

"Are you from the States?" a London cabbie asked me upon my arrival at Heathrow. "You speak English with an American accent."

But it wasn't in Europe that I found my Americanness. I found it in Africa, of all places.

I went to Nairobi, Kenya, in 1987, to visit a friend, Sheila Rule, who was working there as bureau chief for the *New York Times*. In the two weeks traveling throughout Kenya and Harare, Zimbabwe, with Sheila, who is black, I was constantly reminded of my nationality.

Two African men who spotted Sheila and me at an airport in Kisumu, Kenya, said that beyond our speaking English with an American accent, they could tell we weren't African because of the way we carried ourselves.

"You're very demonstrative," one of the men said. "You tend to gesture with your hands, you walk with your heads held high. Most African women don't act that way. Even the way you dress. It's very Western. And we can tell that your American accent isn't fake."

An African man in Harare noticed that when Sheila and I walked into a restaurant with a white male friend, we immediately asked to be seated at a table near the window. "Americans have this sense of entitlement," he explained to us. "That's how I knew you were American."

Africa may be the motherland, but America is my homeland. I realized it then. And I embraced it. I embraced its uniqueness and its freedom. In spite of its faults, it is my home and my native land.

2. Little Things in the School

Why All the Black Kids Sit Together

The black female college student had been raised in an upper-middle-class family—her father, a respected foreign correspondent; her mother, an accountant in a major firm. Much of the student's adolescence had been spent overseas attending schools for the sons and daughters of diplomats. Hers was a privileged life of nannies and housekeepers, and summers filled with travel to exotic places, tennis, and horseback-riding lessons.

The closest she'd come to life in America's inner cities was what she'd read in newspapers or sociology books.

But there she sat—the only black face in a class at the small New England Ivy League college she attended—being stared at by white classmates and a white professor expecting her to answer a question about "gangsta rap."

"How the heck should I know," she said in exasperation. "When rap music hit the American music scene, I was in Nairobi, Kenya."

Her father told me the story. He was so angered by his daughter's treatment at the school, he and his wife were encouraging her to transfer to Spellman, the prestigious black women's college in Atlanta.

"At least there, she won't have to be faced with white students telling her she doesn't 'act black,'" the father said.

In talking with people about this book, I was asked whether I was going to include a chapter about little things that occur in schools.

"You've got to," Knobby, a black D.C. youth asserted. "There have been race riots on school grounds because of the shit white students do to black students."

I asked him to jot down a few of those little things whites do in school. His list contained the following:

1. Expect black students to listen to "white music"—i.e., heavy metal, punk rock—at school social functions.
2. Act like we're antisocial when we don't want to go to their dances or sit next to them in class or with them in the cafeteria.
3. Think all black people are alike.
4. Think they can't be prejudiced, because they're fans of Michael Jordan and Will Smith.
5. Act like black students are dumb if we don't know the answers.

6. Talk to you in class, but never invite you to their houses.
7. Don't mind you talking with white boys or girls, so long as you don't date them.
8. Think that any black student who gets admitted to a good college or university got in under an affirmative-action policy.
9. Act like we can't afford to do some of the things they do, like traveling overseas or paying for a tutor.

The list went on, mentioning white teachers who are quick to place black students—especially black males—in remedial classes or non-college-bound tracks; school administrators who are quick to accept the word of a white student over that of a black student; and counselors who still encourage black students to learn a vocational skill, "so we'll have something to fall back on when we fail," Knobby wrote. "The assumption is you're not going to amount to much," he explained.

I spoke also with a black high school basketball player who swore he could prove that black players were penalized more than whites by game referees, be they white or black. (As a sportswriter I'd heard that same allegation made by other black athletes, in the collegiate and professional ranks. True or not, the feeling was so prevalent that black coaches were encouraged by athletic directors, black and white, to add white players to their rosters and lineups as a way of increasing the number of favorable penalty calls and thus increase their chances of winning.)

A black junior high school student who lives in a rural

upstate New York community said that the school bus driver "never cuts the black kids any slack" if they're late getting down to the pick-up point, but "white students can be standing inside the front door of their homes, keeping warm on a cold day, and you bet the buses will wait."

In April 1999 two white male students at Columbine High School, in Littleton, Colorado, killed eleven of their classmates and a teacher before taking their own lives. Shortly after the incident was broadcast across America, my phone began ringing. On the other end were black friends—parents and students—wondering aloud whether a group of black youths calling themselves the Trench Coat Mafia—the name of the group the two killers belonged to—would have been allowed to attend school wearing black trench coats and toting attitude.

"Think about it. They want to expel our kids from school if they wear gold chains or certain kinds of tennis shoes," said Mrs. Milton, a friend from Washington, D.C., whose son, Christopher Ross, plays tennis on the junior U.S. circuit and has faced more than his share of racial slights. "From what I've read, these two boys threatened other students, brandished weapons, and had been sent to a juvenile center shortly before the incident. Now, think about that. How many black students would have been allowed to exhibit that kind of behavior, especially toward white students, and not find themselves being questioned by the police."

Apparently there were warning signs. Klebold and Harris spent most of the year planning the attack, established a Web site of hate, threatened other classmates, and intimidated neighbors. The local police and school administrators thought nothing would come of it.

"Kids being kids!" After all, these two kids came from white upper-class stock. No need to worry.

Unless, they are black!

In September 1999, five months after the massacre at Columbine High, seven students, all black, were expelled from their high school in Decatur, Illinois, for fighting at a school football game.

No one was seriously hurt. The injuries amounted to bruises, a black eye, a bloody nose. Yet six of the students were arrested and four were charged, as adults, with mob action—a felony—and one was also charged with battery.

School administrators said the two-year expulsions were in line with the district's "zero-tolerance policy." Despite appeals from the students' parents, the school board refused to reduce the duration of the expulsions. The Reverend Jesse Jackson, who managed to successfully free American hostages in Beirut and Iran, went to Decatur to appeal to the school board, bringing with him the glare of media cameras. When school officials refused to budge, Reverend Jackson took his appeal to the public, with mass demonstrations. Illinois Governor George Ryan intervened and managed to convince the school board to show some leniency toward the students.

The board reduced the expulsions to one year, with an option to attend an alternative school for troubled students immediately. The students and parents were expected to accept this gesture of leniency.

On the advice of an attorney the students filed a federal lawsuit against the school district, charging that its zero-tolerance policy was arbitrary and unreasonable. Not to mention racist.

Jackson, being a politician as well, initially said that he

didn't think race was a factor. Later, however, Reverend Jackson said it became apparent to him that the students had been given harsher penalties because they were black. The judge handling the case was told that school officials had examined the students' academic records before deciding to expel them. What did their grades have to do with the fight?

Even though public polls showed that most Americans felt the students deserved some leniency, the judge ruled in favor of the school board and upheld the original two-year expulsions.

Kathy Williams, a lawyer who is a member of my family's church, Tenth Street Baptist, in Washington, believes white institutions—from the courts to the schools—are far more willing to overlook hostile, antisocial acts committed by white youths than they are similar acts by blacks.

"But black kids are always the ones being labeled as 'angry' or 'hostile,'" said Alvin Wright, a black professional and father of two who lives in Houston.

Not to mention emotionally disturbed.

A story that appeared in the December 14, 1999, edition of the *Washington Post* noted that an "African-American male who acts up in class, sasses a teacher or gets into a fight at a Montgomery County school is more likely to be labeled emotionally disturbed than a white student."

And, according to a Board of Education report, a black male student is more likely than an emotionally disturbed white male to be bused to the county's most restrictive school.

Montgomery County is a mixed working-class community in Maryland. During the 1970s scores of black Washingtonians fled to the surrounding suburbs of Maryland and Virginia in search of wide-open spaces and quality education. Suburbs like Montgomery County and Prince Georges County, once working-class white enclaves, became predominantly black and Hispanic communities. The shift in population seemed to trigger a downward spiral in neighborhoods and schools. By the late 1980s Prince Georges County had one of the highest homicide rates in the Washington metropolitan area (a "black thing," some may think), and schools that once were a symbol of progressive education had regressed.

In recent years black parents began to complain about a disparity in treatment between black and white students in Montgomery County schools, only to have their voices fall on deaf ears.

At the time, blacks represented 21 percent of the county's 120,000 students but represented 35 percent of those labeled emotionally disturbed.

The Board of Education's report was done after a group of parents threatened to file a suit against the county.

"You can't get away from the fact that there's a racial issue here," Ray Bryant, director of the county's special education program, told the *Washington Post.* "Is it racist? We have to figure that out. Can you be a benign racist? I don't think so."

I'm not so sure I agree. Take, for example, the brouhaha that erupted at Brother Rice High School in Chicago.

In May 1991 a group of black students at the predominantly white Catholic school for boys decided to hold a separate senior prom. The black students said they resented the fact that their culture did not get the attention that white culture did. In English classes the focus was on the writing of white writers; in history classes, the contributions of black Americans were seldom recognized except during black history month.

And when the white-majority student prom committee hired a rock band and a disc jockey and announced that the prom's music list would be based on suggestions from the majority of the senior-class students, the black kids had had enough.

They knew that the majority would rule, and that meant Bon Jovi over Boyz II Men. That meant slam dancing over electric slides. So the black students decided to have a separate "blacks only" prom and found themselves accused of being antisocial and insensitive to their white classmates.

The prom problem wasn't simply a Brother Rice problem. Other black students voiced similar complaints.

"You go to their parties and they're playing this heavy metal, hard-rock music that nobody can dance to," said Camika Winter, sixteen, a biracial student who was among the peer counselors for the Anti-Defamation League in 1997. "When we complain about the music, they say we're going on some race trip. But when you put on some hip-hop or rap music, everybody, even the white kids, are out on the dance floor. What happened in Chicago with the two separate proms, I can understand how it could come to that. I've been invited by some of the white students to parties, and I just don't go."

If it isn't the music, it's the old familiar dance-partner problem.

Tameeka Robinson, who was a seventeen-year-old student at St. Catharine Academy in the Bronx when she spoke with me, explained. "OK, so you go to an all-white or predominantly white party, and some good soft music comes on. You look around, and there's no black male student there to dance with. Do you think the white boys are going to ask the black girls to dance? No way! So we sit there, looking left out and feeling stupid. And if one of them does ask you to dance, do you do it and chance getting ratted on by your own kind for dancing with 'one of them'? Or having the white girls, who are left standing around, staring you down all jealous and stuff? Rather than deal with that, I just don't go."

My instincts and gut tell me that in none of the examples cited here were the white students being maliciously racist. These were benign racist acts, committed by a generation of whites raised to believe that while blacks should have rights that whites need to respect, the decision of when and where to give that respect rests in the hands of the white majority. What's racist about majority rule?

Were the black students at Duke University engaging in reverse racism when they said that black male students went out with the white girls on campus only "because they were easy"?

I went to Duke in Durham, North Carolina, in January 1993, as part of a monthlong sabbatical granted by my newspaper. One of the classes I enrolled in dealt with race in America. On this particular day, the professor, a young white male, and his teaching assistant, a black doctoral student at Duke, were exploring with the class of about thirty students, most of whom were white, the subject of interracial dating. But it was the handful of black students who were the most vocal on the subject.

"The black males on campus all go out with white girls," said one black female, gathering nods of affirmation from three other black girls clustered near her.

"Why do you think that is?" the professor asked.

"Because they're easy," she said with youthful self-assurance.

"'Easy'?" the professor said with a raised eyebrow.

"Yeah," the student went on. "They'll give it up and the black guys know that!"

Now, I'm sitting there, trying to keep my mouth shut, even though I could have participated in this repartee, waiting for the white students' rebuttal. There was none.

"The only reason white girls go out with black guys is because they feel guilty," said another black female. "They are afraid that if they say no, somebody's going to think they're racist or something."

Still no response from the whites present.

Finally the teacher asked the white students how they felt about what had been said.

"I just want to ask the black guys if I can get some of the names and numbers of these girls they're dating," said a white male student, "'cause I've been at Duke for three years now and I haven't had a date yet!"

There was laughter. Nothing like comic relief to cut the tension. But still the deeper question had not been addressed. I raised my hand, introduced myself to the class, and said that I was taken aback by the sweeping generalizations being made by the black students.

"Ladies, they've all but called you guys—distinguished students at Duke—no less, sluts! How do you feel about that?"

"It's not fair," said one white female. "But if we say something, then we'll be attacked and accused of being a racist or insensitive to blacks. You can't have a discussion when people feel the way she does."

She had a point. Had one of the white girls in the class said she didn't date black men because all they are after is sex or because she didn't find them attractive, for sure, she would have been labeled a racist.

When the class was over, the teacher and his assistant told me that one of the biggest problems with the class was getting the students, especially the white students, to talk about race. It wasn't as though Duke was free of racial tension. I was told by some black students that racial epithets had been scrawled on dormitory walls, that some white professors refused to call on them in class, that they felt unwelcome at white fraternity parties and so didn't go. And yet, in class the white students felt compelled to do and say the right things.

Later I learned that what angered many of the black female students at Duke was that many of their black male coeds were dating white women while few, if any, black women were dating white men on campus. So why didn't they just come right out and say that in class? Maybe they should have been getting on the brothers' cases instead of focusing their attack on their white female classmates.

"We didn't want it to appear that we were criticizing our black brothers in front of the white kids," one black female student later explained.

Leslie Terry, whose almond-shaped brown eyes, high cheekbones, smooth café au lait complexion, and five-foot-nine-inch stature comprise beauty by any standards, said white students at her suburban Maryland school had never voted for a black homecoming or prom queen in the four years she was enrolled there.

"They always choose a blond who looks like that actress on *Buffy the Vampire,*" said Leslie, whose engaging manner and smarts—she had a 3.5 grade-point average—made her one of the school's more popular students. "Some of the black students wanted me to run for homecoming queen my senior year, but I refused. I didn't want to deal with the hassle."

Black students at other schools voiced similar resentment about being shunned and shunted from lead roles in school plays because white teachers and students are reluctant to cast them in *Romeo and Juliet*–type roles opposite white students.

What's up with that?

Would white parents really want their handsome teenage son publicly confessing his love for a black girl, no matter how pretty she might be? Would black parents want their darling black teenage daughter up there on stage kissing some white boy with acne? It may only be a play, but how would it play in Peoria?

Regardless of grade level, black students old enough to recognize racial preference all spoke of "clustering" as a real lighting rod for racial tension—clustering being who sits where in the cafeteria and the school auditorium.

"Black kids in high school, for example, don't want to talk about sensitive racial matters around white students, because they think whites will take their words out of context," said Dr. Andrew Hacker, author of *Two Nations: Black and White, Separate, Hostile, Unequal.*

Rachel Weiss, who served as a youth counselor for the

Anti-Defamation League, said she'd noticed something at her integrated high school in Brooklyn.

"For years the twirlers and cheerleaders had all been white, and the booster squad had always been black. About five years ago black girls went out for cheerleading, and now it's more diverse. But twirlers were still all white until this one black girl tried out and made the squad, and this year she's the co-captain. Now a lot more black girls are coming out for twirlers. Sometimes it takes one strong person to try it, whether it's the twirlers or the lunch table."

Max Weisman, another counselor, who describes his heritage as Catholic and Jewish, said the cafeteria scene at his high school in Rockland County, New York, was like a demilitarized zone.

"There was a Haitian table over there, a white table, a black, and a Hispanic table. I think everybody does it to a certain extent. I don't do it on the basis of race. It's, like, who you feel more comfortable with. But we have to ask ourselves why we feel more comfortable with certain people than others. And race shouldn't factor into that."

What annoys many black students is the feeling they are being blamed for something that white kids themselves do.

"Hey, they all sit together and nobody says anything," said Kim G. of Washington. "It's only a problem when *we* do it. Besides, why do we have to break the ice? They can come over here and sit with us, if they really wanted to."

Thomas Kochman, a diversity consultant and white male who has dealt with the issue of "clustering in the workplace," calls this the "woo-shoo" phenomenon.

"Who's doing the wooing and who is doing the shooing," he explained. "As long as whites could do the shooing they were comfortable. Now that blacks are doing the shooing, whites are feeling uncomfortable."

Growing up in the 1950s, I used to think all white kids were smart. They had to be. Their parents controlled the country—ran all the major businesses and the government. White kids had better schools than black kids, and, we were told, smarter teachers. They got the new classroom textbooks; we got their hand-me-downs, replete with their names and personal markings. I'd always wanted a textbook in which my name appeared first.

As the youngest of seven children, I'd had my share of hand-me-downs from my older siblings. Now I was getting secondhand textbooks defaced by little white boys and girls with Jack-and-Jill names, and I hated it.

So I believed in my heart, mind, and soul that all this made white kids smarter. If they weren't smarter than black kids, the Supreme Court wouldn't have declared separate-but-equal schools unconstitutional. In my adolescent mind, that was declared because they knew white education was superior to black. By simply sitting next to white students in class, some of those smarts were bound to rub off on underprivileged black children like me.

As life would have it I never attended an integrated public school. And I do not feel at a loss for not having done so. In my little black classrooms, I learned more than the basic three R's—reading, writing, and arithmetic. I learned how to think and do for myself. I learned that the color of my skin had nothing to do with my ability. I learned to be thankful for a free education, an opportunity denied my ancestors. And I learned to make the most of that opportunity. I was taught these things by black teachers, who went out of their way to encourage their students.

But our teachers also never allowed us to lose sight of the facts: America was a racist society. Blacks were still considered to be inferior to whites, and no matter how intelligent we might be, we would always have to work twice as hard to get half as far. Now, that may sound like a cruel thing to say to young impressionable minds. But even as children, I and many other blacks knew our teachers' words were meant to help—not harm—us, to prepare us for life as a "minority."

My first integrated educational experience did not come until 1972 when I enrolled in Columbia University's Graduate School of Journalism. The class—a mix of races and ethnicities—had its share of know-it-alls. But never once did I think the white students were any more intelligent than I or other blacks in the class were. We all struggled to meet deadlines at one time or another. Poorly researched term papers weren't a race-specific phenomenon. In fact, I found the black students to be more self-confident and far more independent in their thinking than most of my white classmates.

What Columbia did for me, and probably for other black journalists who attended the university, was provide a kind of legitimacy not received from historically black undergraduate schools. With all due respect to Howard University, my undergraduate alma mater, I sometimes get the distinct impression when moving about the white world that my credentials as a journalist do not have sufficient merit until I mention my masters degree from Columbia.

I attended all-black schools because I had no other choice. My sisters have chosen to send their children to all-black schools because they believe the children will be treated more fairly than in predominantly white schools.

Forty-six years after the Supreme Court struck down this country's apartheid system of education, schools are as segregated as ever.

Integration was supposed to diversify our classrooms. It didn't work out that way. The minute mostly white schools became a little too black for comfort, white parents transferred their kids to private schools (Negroes need not apply) or moved to suburbs where the racial makeup more or less assured them their kids would be attending school with their own kind.

Despite the best of intentions the fact remains that the nation's three largest public school systems—Los Angeles, New York, and Chicago—are predominantly black and Hispanic. Black teachers and administrators are still assigned to work in predominantly black schools. Disparities still exist between the quality of education received by white children and that received by black children.

We know it, and there's not a damn thing we can do about it. No, really; I mean it. This is a free country, folks. Forget that crap about children having to go to the school within their district. If we don't like the school district, we can move to another one. If we choose to stay, we can enroll our child in private school.

Americans are a defiant lot. We do not like being told—by the government or the courts—what we have to do, especially when it comes to our children. The courts have even thrown up their hands and all but declared the issue of school desegregation closed. We tried; we failed; might as well get over it.

Really. I'm not a pessimist, folks. I'm a realist who is open to suggestions.

3. Little Things in the Home

Did a Black Person Own This House?

Dr. Mel Rapelyea, a black radiologist, remembered being told by his white real-estate broker, who was also his friend, that if he wanted to sell his English Tudor house in Columbia, Maryland, a popular upper-class integrated suburb, he'd better remove all the "black stuff" from the house.

"And put some Pillsbury chocolate chip cookies in the oven," George Clark, another white friend, jokingly told Dr. Mel, as he's commonly known. "It works every time."

The three were laughing about the matter over drinks during the Memorial Day weekend in May 1998. The whites in this mixed social gathering showed they were well aware of the problems many blacks face when trying to sell their homes in a racially biased society that all too often plays to stereotypes.

"I can't sell my house because whites don't want to buy it," said Dr. Mel, who had two solid offers, one from an Indian couple, the other from a white couple that was later withdrawn. "I didn't want to believe that [racial] issue was important."

Larry, white and another weekend guest and friend of Dr. Mel's, wondered if race was really a factor. "If Mel had lime green carpet in his living room, would someone walk in and say 'Oh man, that's not attractive to me. I can't see myself living here.' "

Mrs. Cregar, a white woman who doesn't mince words, answered for Mel. "You can say, 'I can take the lime green carpet out and I can put in beige,' but you can't say, " 'A black person lived in this house and I don't want to be in this room. I wouldn't want to put my kids to bed in the same room where a black person slept.' "

Dr. Mel, shaking his head at the absurdity of the situation, added, "But I can have a white person on my angio table dying and I'm trying to do something to save his life, but that person won't buy my house."

A white New York real-estate broker who was asked about the role race plays in the housing market answered on the condition she not be identified by name so as not to compromise her professional relationship with potential buyers. "One woman told me she didn't want her children sleeping in a room where blacks had slept. Another said that homes owned by blacks smelled funny."

Her advice to someone in Dr. Rapelyea's position was to remove all items from the house that could readily be associated with a black lifestyle. In other words, ditch the African art, tuck the family photo album in a drawer, and throw out the old copies of *Ebony, Jet,* and *Essence.* Now toss a few copies of *Architectural Digest* and the *New Republic* around the place. Tack up a photograph of Ronald Reagan and a poster of the Hanson Brothers. And make sure the Shaker furniture is prominently displayed in the living room. Now, that's a race-neutralized house!

Although blacks and whites may work together and share the occasional drink after work, America remains a largely segregated society, especially when it comes to the communities in which we live. Blacks move in, whites move out or farther out.

In his book *Two Nations,* Andrew Hacker—who, I remind you, is white—wrote honestly and openly about "black contamination."

"Americans have extraordinarily sensitive antennae for coloration of neighborhoods. In virtually every metropolitan area, white householders can rank each enclave by the racial makeup of the residents. If you are black, these white reactions brand you as a carrier of contaminations. No matter what your talents or attainments, you are seen as infecting a neighborhood simply because of your race. This is the ultimate insult of segregation."

Odessa Woolfolk, an educator, lecturer, and former director of the Civil Rights Institute in Birmingham, Alabama, says she illustrates with a simple exercise in some of her lectures on race the schism in American housing patterns. "I ask those present to jot down the race of the people who live directly across the street from them, then list the race of the person who lives next door to them, and the race of those who live at least two doors from them," she said. "Inevitably, most whites end up with all whites on their list, and most of the blacks have only blacks on their list. There are the occasional exceptions, but for the most part, that has been the rule."

There are whites who have never set foot inside a black person's home and blacks who have never been inside white homes, other than to work there. Consequently, we see only what we want each other to see.

Our homes reflect our tastes, attitudes, beliefs, desires, and heritage. And yes, you often can look at a home's contents and tell the color of its owners.

I looked around my Manhattan apartment. There is a predominance of Afrocentrism. African art on the walls, African sculpture in the foyer, African mud-cloth throw pillows, a Romare Bearden print, photographs of Langston Hughes and Malcolm X. Except for a black-and-white photograph of John F. Kennedy and a poster from the Sundance Institute, you wouldn't have to be Sherlock Holmes to figure out a black person lives there.

I've been in homes decorated with an eclectic mix of art, yet was quickly able to ascertain that a white person lived there, if for no other reason than its furnishings were void of race or ethnicity.

We can't tell each other how to live. But we are the images we live with. We surround ourselves with things that make us comfortable.

Imagine how uncomfortable it must have been to Pauline Schneider, a Washington attorney, and her husband, Roy, in the 1970s when they and their two small children moved into a San Francisco house previously owned by a white woman who was an avid movie buff. The woman had decorated the kitchen floor with old movie posters and covered it with a clear floor shellac. "It was kind of odd but interesting," Ms. Schneider recalled thinking. Until the day her then four-year-old daughter, Suzanne, now a woman in her twenties, burst into tears at the dinner table.

"Mommy, I don't want to be black." When asked why, she said, "Nobody likes black people!" Asked why she would say something like that, she said, "Because there are no black people on the floor." Pauline told me, "I hadn't noticed that only white people were in the posters, but here was my four-year-old child—whose mommy was a lawyer and daddy a doctor—getting this message. Our whole thing is that our children shouldn't have to bear the burden we had."

It probably never occurred to the previous owner that a black child looking at her floor of fame would see not a tribute to Hollywood's great movie stars but a person who didn't like blacks.

Still, we decorate our homes without ever once giving thought to ethnic diversity. Each time a white American places a piece of African art in his or her home, he or she is showing an appreciation for African/black culture. Blacks do the same when they hang artwork created by white artists. Everything—from the books we read to the television shows we watch—conveys racial preference. When my adolescent nieces come to my apartment, they see photographs of friends who are important to me, black and white in equal numbers. The message I am conveying is race sometimes doesn't matter.

What surprised me throughout my various focus-group sessions were the numbers of blacks and whites who said they had members of the opposite race whom they considered close friends. So I decided to try the following exercise with a mixed-race focus group in Washington.

I asked the twenty people present, how many had close friends of the other race. Nearly every hand was raised. I then asked them to keep their hands raised if they'd ever been to their close friend's house for dinner. No more than ten hands were still in the air. Then I asked how many had spent the night at their close friend's home. Three hands, mine included, remained raised.

"What's your point, Lena?" Roslyn Frazier, who is black, wanted to know.

"My point is simply this. We have formed stereotypes of each other based largely on perception, a perception gleaned from a few hours spent each day working or learning with members of another race."

How would white people know what black people's homes smell or look like if they've never been to a black person's house? Having worked as domestics, blacks may be in a better position to know what goes on behind the closed doors of many white Americans. Yet fewer blacks than you'd think have ever been invited to sleep over at a white person's home.

Stereotypes begin at home. They seldom remain there.

"Aunt Lena, you're dancing like a white girl," my ten-year-old niece, Lauren, said to me one New Year's Day when members of the Williams family had gathered at my brother's house for good food and good times.

"How many white girls have you gone dancing with?" I asked as she giggled.

Where had she heard such a thing? Her older cousins, perhaps?

"I see them dancing on TV and during halftime at the games," Lauren explained.

"And that makes you an authority?" I replied. We all laughed the matter off. But, if nothing more, perhaps I had raised enough doubt in her mind to make her think twice.

Home is a safe haven. A refuge from the outside world. A place where we are free to be ourselves, speak our minds freely, and do as we please. When I'm at home with my family I sometimes find myself doing and saying things I might not do in the presence of outsiders, black or white. Little things. Like... well, like counting the black players on sports teams.

It's something my parents used to do. Whenever we sat down to watch a sporting event on television, we'd count the number of "colored" players on the teams. We cheered for whichever team had the most.

Across America, black families, in addition to my own, often choose sides based on the number of black athletes playing for a given team. While whites may find this childish and petty, we see a historical significance in our allegiances.

"For a long time the Washington Redskins didn't have a black player on their team," my brother Ralph recalled. "Although we wanted to root for the home team, blacks had a problem with rooting for a team that refused to allow blacks to play, even after other teams in the National Football League were recruiting black players. So yes, we've always counted where it counts."

Sports was, and in some respects still is, the one arena in which blacks were allowed to excel. To show our prowess. Our superiority. A victory on the athletic field was viewed as a victory for the race.

Joe Louis. Jesse Owens. Jackie Robinson. Wilma Rudolph. These are our heroes and heroines. Men and women whose victories in the ring, on the field, and on the track did the race proud. They seemed to make us "one nation under God," their triumphs cheered by blacks as well as whites.

One of my most vivid recollections as a child was being told of the Joe Louis–Max Schmeling fight. White Americans felt Louis was a credit to his race and wasn't uppity like "that nigger Jack Johnson."

That night, black families across America gathered around the radio to listen to the fight. When Louis won, one of our neighbors ran through the streets like a black Paul Revere sans horse, shouting the victory to those who might not have heard the news. And I remember thinking, *This Joe Louis is really a special man.*

Although blacks have made significant gains in sports as participants, there is still considerable controversy over the lack of blacks in front-office positions as well as in coaching and managerial positions. Regarding such, some of my people, my family included, have set new measures for team loyalty: Does the team have a black coach? Does the team have a black quarterback?

Truth be told, this kind of racial allegiance can create conflicts.

Imagine my surprise during the 1997 NFL championship game between the Tampa Bay Buccaneers and the Green Bay Packers. For as long as I can remember, our family had rooted for the Packers. Not only were there several black players on the team, many played key positions. Vince Lombardi, the legendary Packers head coach, was described as a fair man. A man who respected hard work. Race didn't matter, so long as you did your job.

But that cold December day, the Williamses had gathered around the television and began cheering for a Tampa Bay upset.

"I thought we were for Green Bay," I said.

"Well, I like Green Bay, but I'd really like to see a black coach get to the Super Bowl, so I'm cheering for Tampa Bay because they have Tony Dungy, a black head coach," my sister Ada explained.

I accepted her logic and joined in the fanfare for the Bucs. We did not suffer greatly when the Bucs lost, because they lost to a team that had forever proven itself worthy of our allegiance.

As a sportswriter I have to uphold the ideal of objectivity in my reporting. What I do in private, however, is another matter. And I must admit that there are still times when, try as I might, I find myself wondering whether this or that call or penalty was just, or just racist.

In December 1999 Dungy's Bucs once again played for the NFL championship, this time against another favorite, the St. Louis Rams. I was pulling for the Bucs, again in the hopes of seeing a black coach in the Super Bowl. To make things even more interesting, the Tennessee Titans, led by Steve McNair, a black quarterback, had already clinched a play-off berth in the Super Bowl. The Bucs' offense was lead by another black quarterback, Shaun King. A black man's and woman's dream come true. Not only did we have a chance of having a black coach in the Super Bowl, we might also have two black quarterbacks! A chance to debunk several myths in one glorifying moment.

Unfortunately, the Bucs lost the game by five points. In the closing minute of the fourth quarter, there was a controversial call by a white official that went against the Bucs, who were on the Rams' twenty-eight-yard line. It was one of those calls that, in sports jargon, "could have gone either way." But that wasn't how a lot of the black folks I know saw it.

"Did you believe that call?" a black colleague whispered in my ear the following day. "He caught that ball. But you know they didn't want to see the Bucs and Titans playing in the Super Bowl."

We may not have known who "they" were, but we knew they were white. (It's a black thing.)

Truthfully speaking, neither one of us felt this penalty call was a planned conspiracy. In sports, momentum shifts and things unfold so quickly, it's hard to predict an outcome. Penalties usually even out, we're told. Convincing us of that is another story.

"Don't ever think that we're the only ones counting players," said my colleague. "Whites do it, too. I attended a predominantly white school in Ohio, and the white students would always pull for the basketball teams from white colleges over those from black colleges. Call it human nature, but whites are as guilty of it as blacks."

At its annual meeting two months later, the NFL ruled that a pass reception like the one caught by the Tampa Bay receiver would be considered valid in the future.

Make Yourself at Home

A black person or couple invites a white friend, colleague, or neighbor over. The white person shows up with children in tow. It's bad enough that the black host or hostess didn't know the white invitees were going to bring the children, but the precious little "Johnnie" and "Jane" walk into the house like they own the place.

They turn on the television set, ask for something to drink, take part in adult conversations, and make demands that would leave some adults blushing.

This kind of behavior places the black host or hostess in the awkward position of having to bite his or her tongue—we wouldn't want the white neighbors to think we're hostile now, would we?—and let kids be kids.

The notion that children should be seen and not heard may not be fashionable in some circles, but where I come from, the rule still applies. Most blacks I know don't subscribe to Dr. Benjamin Spock's way of raising kids. It may be fine for white folks to let their children run and roam free, to spare the rod and spoil the child, to indulge their every fancy. But black parents know better. Even black parents who can afford to buy their children whatever their little hearts desire, often don't because they want to show them that in a race-conscious world, you can't always get what you want. Better to learn that lesson at an early age from one's loving parents than to years later learn the hard way, when you're passed over for a job promotion given to a white coworker you trained to do the job.

Black parents raise their children to face the world as though they are preparing for battle, throwing around them a protective armor of wisdom to steel them against the harsh reality of the racism the children are sure to face sooner or later.

"You can listen to that white folks' foolishness all you want; I'm going to raise you the way I see fit," my mother used to say, echoing a sentiment heard across black America.

Coming of age, I wanted the kind of freedom I saw on

television shows like *The Brady Bunch* and *The Partridge Family,* where young adults were allowed to make mistakes and learn from them. How come black kids weren't allowed to roam free, neck in cars in secluded parks, hang out at drive-in burger joints? (And yeah, I'm showing my age here, OK?) Why couldn't my mother be more like Donna Reed? Why did I have the sneaking suspicion that, unlike Robert Young, my father didn't always know best.

My mother and father, like so many other black parents, knew better. They knew that white authority—the police, in particular—were not going to give black kids the same breaks or benefits of doubt given white kids. My mother made sure her children didn't venture far from home and always traveled with a group. She believed in the old adage of there being strength in numbers. We were taught how to safely navigate two nations: one black, one white, separate and unequal.

"Showing out in public" was not to be tolerated at any age where I came from.

In supermarkets I've seen white parents try to explain to their screaming children why they can't have another toy or more candy.

"Bobby, you already have twenty action characters; you don't need another one." Or, "Amber, too much candy isn't good for your teeth; you know what the dentist says."

All this reasoning, while little Bobby or little Amber is rolling around on the supermarket floor for all to see.

My sister Ada, the mother of Lauren, eleven, believes white parents may act this way publicly to demonstrate to the world "what loving, tolerant parents they are, unlike us black parents, who are prone to let our kids know—in loud, stern voices—*no* means *no,* and give them a whack on the butt or hand if they continue to show out."

Many a black child's rear end, this one's included, was spanked for disobeying orders, and we truly believed our parents when they said that the spanking was hurting them more than it hurt us. It was that way in the 1950s, and in some respects, it's the same today.

I was raised to mind my manners—what we liked to call basic home training.

In no way is that a black thing. There are blacks who are ill-mannered and whites who are ill-mannered. Yet white parents appear to be more willing to tolerate or ignore lapses of etiquette and manners committed by their children. Black parents, on the other hand, tend to demand respect. Understand why: We as a people get so little respect in the outside world, we're not about to be disrespected in our homes.

Blacks as well as whites suggested in the focus groups that some of the little things I mention here have less to do with race than with a lack of basic home training.

A black man in Chicago told me that he really gets annoyed when his white boss waltzes into the office in the morning and starts barking out orders and requests without so much as a "Hello," "Good morning," or "How are you?" Honey, that ain't about race. That white boy lacks home training. And the same would apply if the races were reversed.

A doorman who works at a stately Upper West Side apartment building said he'd noticed, over his twenty years there, that the few black residents in the building—adults and children—were more likely to say "please" when making a request of him and "thank you" for his services than were whites in the building.

"Some of the white kids here act like we're being paid to work for *them,*" said the doorman, who is Hispanic. "One of them actually said so—that his parents gave me a nice tip at Christmas, so why was I giving him grief about watching his double-parked car."

Was this an example of white privilege or a kid lacking home training? Take your pick.

Sometimes we think certain acts or actions are unique or distinct to one race, only to find out our assumptions are wrong. I discovered that during my weekend in Betterton.

I was talking about how black parents know how to put the fear of God in their children, with nothing more than a simple look. If that doesn't work, there is always the pinch, most often used in public places, like churches, where parents or grandparents exercise discretion and subtlety in their discipline.

"Hey, that's not just a black thing," said Francine Cregar. "My mother would give you a nice little pinch, right under here." She demonstrated by reaching for the fleshy underside of my upper arm.

How'd she know that? By God, that was the spot.

"And all the time, I thought that was a black thing," I said as I tried to wrest my arm from her fingertips.

"No, honey, that's a mother thing," Francine retorted.

Kids will be kids, race notwithstanding. But by the time you're in your teens or twenties, you should know better. It's fine for young white college students to "have a little fun" during spring break: skinny-dipping, guzzling beer, and wreaking havoc on public streets. "It's just young people who need some release from the stress of college life," you say. *Right!* Let some black kids show up in town and start tearing off their clothes in public, drinking, and hanging out till dawn, and it's time to bring in the National Guard.

For years, black college students converged on Atlanta, Georgia, during spring break for the annual "Freaknik" celebration, which attracts thousands. There was the usual drinking, partying, fooling around between the sexes. But in 1995, white residents and white businesses complained that the crowds from Freaknik were disrupting traffic and hurting the flow of business into their stores. They convinced Mayor Bill Campbell, who is black, to clamp down on the celebrations that dated back to the early 1980s. When the City Council withdrew permits for concerts and other events in public parks for the 1995 Freaknik celebration, the students staged a protest, which turned into a confrontation with arrests being made.

The Freaknik celebration, the black students argued, was no worse, in size or scope, than the spring break celebration that annually took place in Florida where thousands of white students party, drink, and sometimes run naked through the streets.

Another rather interesting racial difference revolves around privacy in the home.

I'm not a parent. But many black parents have told me that when their children start paying rent, the children can have all the privacy they want. Until then it's an open-door policy.

I find that many white parents support a more liberal concept that parents should—to a degree—respect their children's privacy, especially after the children enter their teenage years.

I've visited white homes where children as young as ten, have DO NOT ENTER or KNOCK BEFORE ENTERING posted on the outsides of their doors, and I've been told by their parents that they always knock before entering their children's rooms. I've been in white homes where kids as young as twelve have private telephones or personal computers in their rooms. Little Johnnie and Jane are locked away in their rooms, with the doors closed, communicating with perfect strangers on the Internet.

What's up with that?

One question that emerged in the aftermath of the shootings at Columbine High School: Why didn't the parents of Klebold and Harris know what was going on under their own roofs?

These two boys were able to spread hatred across the Internet, make bombs in the garage of one of their homes, and hide weapons in plain sight in their dressers, and yet their parents didn't know what was going on.

"I really believe that whites are too arrogant to see the error of their ways or their children's," David Crandall, my former brother-in-law told me in the fall of 1998, prior to his death. "I used to volunteer at a youth center in Virginia, and whenever I'd tell a white parent that their kid did something bad, they always questioned or challenged me. 'Are you sure? Because he's never done that around me.' But when I talked to a black parent about something their kid was doing, they'd listen, and even if they didn't believe what I was saying, they wouldn't say it in front of me and the child."

"It's not a matter of trust," said Clarice Williams, my sister-in-law. "I always trusted Angel. And I gave her a lot of privileges. But kids will be kids. Behind closed doors they get into all kinds of devilment. And the older they get, the more likely they are to test bounds and limitations. In order to be on top of things, you need to know what's going on. So yes, I did impose certain restrictions on her and made sure I kept tabs on her."

For a generation raised in an integrated America, these little things must seem petty and silly. There are spoiled, rotten black kids; black kids whose parents indulge them; black teens who appear to have the run of their households. Some blacks blame integration. Those of us who had been denied did not want to deny our children. More than ever, we're raising a generation of black kids who often have to be reminded they're black.

A sobering example of the difference a generation makes when it comes to race matters was illustrated in the way Carolyn Kennard, another participant in the Houston focus group, and her twenty-year-old son interpreted the racial dynamics of an encounter in a Houston mall.

Her son chose to shrug off the incident in which a saleswoman in the store he'd just visited completely ignored his attempts to purchase a shirt. Ms. Kennard said she would not have taken the path chosen by her son—to go to another department store—but would have immediately complained to the saleswoman's manager.

"Mom, what good would that have done?" her son wanted to know.

"If nothing more, it would have made me feel a whole lot better," Ms. Kennard said with a satisfactory smile.

Often young blacks raised in nearly exclusively white environments suffer from bouts of racial amnesia: They forget they're black. Black parents like Mazie Williams; Dr. Lorna McFarland, a pediatrician, and her husband, Kenneth Noble; and Dr. Manley know this and struggle every day to prepare their children—raised in integrated neighborhoods and educated at predominantly white schools—to understand who they are, because whites won't allow them to forget. Stories abound about black children who live the life of Oreos—black on the outside, white on the inside—only to be painfully reminded of their blackness.

During an all-black focus group in Houston, Texas, in June 1998, Mazie Williams, the mother of a thirty-year-old son and eighteen-year-old daughter, said that she has always reminded them that they're black.

"You may say, 'Why do you have to remind them that they're black?'" Mrs. Williams said, cutting me off at the pass. "Because I don't want them to get so complacent that they think everything is rosy. They're still going to have to work extra hard to get equal dollars."

Mazie knew that her children lived in a predominantly white world and often moved in white circles. Doing such, they sometimes let down their guard, forgetting they were black.

She mentioned one incident in particular to illustrate her point.

"My son and a white woman he had been dating for four years were on their way to a movie and stopped at a service station to purchase gas. While my son was pumping gas, the girl went into the station to give the attendant the money. A white police officer who was standing in the station looks at my son's girlfriend and said she looked familiar and asked her name. She told him and then walked back to the car."

Mazie said the girl noticed the officer was still watching her and saw her get into the car with Mazie's son.

"They had no sooner got in the car and pulled out onto Richmond, when they saw the officer in his cruiser, with his lights on," Mazie went on, her voice growing shriller with each memory of the incident. "He pulled them over and asked for my son's license. When he ran my son's license plates through the computer, he found out that my son had an outstanding ticket. The next thing my son knows, he's being handcuffed and taken off to jail. The girl was allowed to leave."

To this day Mazie is convinced her son was stopped for nothing more than being with a white woman.

Don Payne, who handles communications for the Houston police department, said the moral to the story—one he abides by and encourages other blacks to likewise do—is to "pay traffic tickets as soon as you get them."

What's up with that?

"Listen, I don't want to be arrested," explained Don, who acknowledged that he had been stopped by police for what he suspected was "DWB": driving while black. "So I know to pay my tickets whenever I get them. My dad always told me, when you get stopped by a police officer, it's 'Yes sir,' 'Yes ma'am.' Always be very courteous, no matter what happens. There are some black folks who think they have to have this African warrior posture. I listened to my father, and for the most part, I've walked away no worse for the wear. Like the time in Chicago I was driving with an expired sticker, expired driver license, and was over the legal limit for alcohol. But when I was pulled over, it was 'Yes sir,' 'No sir,' and the officer gave me my tickets and let me hail a cab home."

Don also said that, had the police impounded Mazie's son's car, the parents of his white girlfriend might have viewed it as the consequences whites pay for dating blacks.

Some black parents, whose children now have children of their own, still recall the year, if not the month, the invitations for sleepovers at their kids' white friends' homes stopped. Other parents still harbor resentment over the times their children were questioned—more like badgered—by their white friends' parents about what their parents did for a living; if they had siblings and if so, how many; whether their parents were college educated. Black parents have also been made acutely aware of instances where white parents demanded to know how their kids knew this or that black youth.

"My son was enrolled in a boarding school in Massachusetts, and he'd always had white roommates because so few blacks attended the school," said a middle-aged black male financial analyst who lives in Connecticut and works on Wall Street. "His junior year, he was rooming with a white kid from Shaker Heights, Ohio. They got along well enough. The kid even invited my son to spend the Thanksgiving holiday with him. Then one day my son said that his roommate and a group of other whites students were in their room in the dormitory, laughing and joking around, and when my son left the room, he heard his roommate say something about 'a nigger,' and all the kids laughed. Do you know, to this day, I can't convince my son that they were probably talking about him? He swears they weren't. My question was, if that were the case, why did they wait till he left the room? Maybe they routinely talk about niggers and didn't want to talk about niggers around him. Either way, it bothered me that he acted so nonchalant about the matter."

Explaining racism isn't easy, especially to children.

Lorna McFarland has traveled the world with her husband and sons, Eric and David. For five years, Kenneth Noble was the *Times* bureau chief in Abidjan. He and his wife now live in an upper-class neighborhood in West Los Angeles. Their sons are enrolled in the elite Elysée School. The Nobles are raising their sons to deal with race by refusing to respond to what they view as "other people's psychoses."

"This is something Ken and I talk about all the time," says Lorna. "We decided that our sons weren't going to hear from our lips the problems and psychosis of prejudiced people. What we've done is emphasize the history, not only of black people but all people—because my children need to understand that black people aren't the only people on this planet who have been hated or despised or placed outside of the accepted realms of society."

On the other hand, Dr. Manley has chosen a kind of hands-off approach, opting instead for what he calls the "natural consciousness process" to set in.

"I saw it happen with my fifteen-year-old daughter Geneva. She grew up in an integrated neighborhood, where she was fond of whites, then probably realized that to maintain a balance between her own black identity and the integrated reality in which she lived, she had to destroy some confidence in her black identity. She couldn't just be all white, because it would negate her."

Dr. Manley said he never talked to his daughter directly about this consciousness shift, but made several innuendos about the changes he saw in her.

"I think she realized that her white friends weren't willing to come along and try to be black, and those who did, did so only up to a certain point. They didn't have a point for her; it was either all the way or no way. Whereas her black friends said, 'To hell with it. We don't have to be with them.' I think she saw that and it helped her restore her balance. But I still think she doesn't recognize that racism had something to do with that."

One way to keep a child's race in perspective is by surrounding the child with positive black images in the home and beyond.

My sister Ada, for example, has this thing about dolls. She doesn't buy white dolls for Lauren. If she can't find a black version of the doll of the year, she won't buy it. And she's not alone on this.

When Cabbage Patch Kids were all the rage, Ada combed the Washington metropolitan area, stopping in every Toys "R" Us store within a fifty-mile radius, searching for a black Cabbage Patch doll. She eventually found one in time for Christmas, but it took considerable effort on her part. She doesn't buy Barbie, she buys Christine, Barbie's black friend.

"How many white kids do you see hugging on black dolls?" Ada asked by way of explanation. "But you go out in public and you see all these little black girls—toddlers and adolescents—lugging around these white dolls with blond hair.

"I want my child to have dolls that look like her and her family members," she went on. "The type of doll you buy for a child sends a message to that child. I'm not talking about the Muppets; they're furry, colorful toys that aren't based on human form. Barbie is another matter. She projects this unrealistic ideal, not just for black girls, but white girls, too. I don't want Lauren going around thinking she's going to grow up and look like Barbie."

Sound silly? Well, think again.

In the mid-1950s when Thurgood Marshall, then a lawyer for the National Association for the Advancement of Colored People, was arguing the *Brown* v. *Board of Education* case before the Supreme Court, he asked Dr. Kenneth Clark, the renowned black sociologist, to conduct a test with black children and dolls. Dr. Clark made it simple enough. He showed the children identical dolls in a black and a white version and asked them to choose the dolls they preferred. An overwhelming majority of black children chose white dolls. When asked why, most of the kids said they thought the white dolls were prettier, even though they were identical to the black dolls except for being painted white.

Dr. Clark's dolls test was cited by the Supreme Court in its decision declaring separate-but-equal schools unconstitutional.

Yet nearly five decades later, many toy stores do not stock black dolls in large enough supply to accommodate the demand. Toy manufacturers are still flooding the market with dolls that reflect white standards of beauty.

"You search high and low to find a black doll, and when you do, it usually has straight, long, flowing hair, and a thin nose," my sister told me with a sigh of resignation. "What are you going to do?"

The Wrong PO Box

Many blacks are raised to believe we don't have a right to anything, except taxes and death.

Whites, however, seem to be raised with an innate sense of entitlement, as if they have an inalienable right to do as they please—"It's our world; the rest of you are just passing through"—and they've passed it down to their children, like precious family heirlooms.

I see this kind of white privilege often—when I ride the bus and a white mother is engaged in a sing-along with her toddler, seemingly oblivious to others on the bus, who might simply want to get to their destinations in peace. I hear it whenever whites question blacks about personal details, expecting an answer because they feel entitled to one. I've lived it when white Americans expect to get a promotion or a last admission slot into an exclusive school or college and blame their failures to do so on reverse discrimination.

A friend who lives in Oak Park, Illinois, telephoned me one day last fall to read to me from Oak Park's weekly newspaper an article that talked about white privilege in all its various forms.

Written by Dick Haley, the publisher—who happens to be white—the column was in response to a lawsuit filed by a black Chicago police officer, who lives in Oak Park, accusing the Shell Oil Company of racial discrimination.

According to reports in the newspaper, the officer had noticed a pattern at the Shell gas station he patronized in Oak Park. Although above all the gas pumps there were signs ordering customers to pay first, white patrons routinely pumped before paying. Anyone who's ever gone to a self-service station knows that the cashier controls the mechanism that triggers the release to the gas pump. What the black policeman realized was that the cashiers, many of whom were white, were bypassing the rule and allowing whites to pump without paying up front, refusing to give blacks the same courtesy.

The policeman was so incensed by the pattern, he decided to videotape the practice. The videotape confirmed his suspicion and he filed suit against Shell Oil.

Mr. Haley, also an Oak Park resident, said in his column that he takes it for granted that the signs are there just as reminders for people to remember to pay.

"I automatically begin pumping first and if the clerk refuses to release the fuel pump, I just give a long stare and it's done," wrote Mr. Haley. "It's part of the privilege that I never associated with color."

Who, for example, would ever think that the sample products we periodically receive in the mail are a form of white privilege?

My sister Barbara Williams Turner and her husband, Carl, both work as mail carriers for the post office in Bethesda, Maryland, an upper-class predominantly white suburb. They've noticed an unsettling pattern "that a lot of the sample products distributed through the mail aren't sent to black neighborhoods," said Barbara as Carl nodded in agreement. "We've seen free samples of detergent, soap, shampoo, cereal, potato chips, and sanitary napkins, among other items, zoned to neighborhoods that are either all white or predominantly white, sometimes two samples to a single household, and not one sample addressed to a known black area."

Carl added, "Sometimes there are boxes of sample items left at the end of our shifts, either because the people they were addressed to had moved or the wrong address was on them." He and my sister admit they've been tempted to take a batch of these leftover samples and distribute them to friends, relatives, and neighbors in their integrated Silver Spring, Maryland, apartment complex. But to do so could result in dismissal, so the items are returned to the senders.

Why should the racial dividing lines dictate such mundane things as free samples or payment procedures at gas stations? More than once I've seen white motorists fill their gas tanks, then drive off without paying, like it was simply an oversight. So dishonesty isn't exactly a race-specific trait. And seeing as how black Americans spend billions of dollars annually on products and services, one would think we'd be prime targets for marketeers to sample their wares with us and whet our appetites. But we don't get the freebies. We're going to have to pay for anything we get, and, in some cases, at markups twice those in white areas, as numerous studies have shown.

Robert Jensen, a white professor at the University of Texas, says white privilege is "the dirty secret that we white people carry around with us every day" and that in a world of white privilege "some of what we have is unearned."

Writing in the *Baltimore Sun* newspaper, professor Jensen noted that white privilege means that if he sought admission to a university, applied for a job, or went apartment hunting, almost all the people who evaluate him look like him in that they are white. "They see in me a reflection of themselves—and in a racist world, that is an advantage. I smile. I am white. I am one of them. I am not dangerous. Even when I voice critical opinions, I am cut some slack. After all, I'm white."

It is because of white privilege, Andrew Hacker asserts in *Two Nations* that "no white American, including those who insist that opportunities exist for persons of every race, would change places with even the most successful black American."

"All white Americans realize that their skin comprises an inestimable asset," he writes. "It opens doors and facilitates freedom of movement. It serves as a shield from insult and harassment."

White skin brings with it a birthright of status and privilege.

Often I am asked whether I believe I am discriminated against more because I am black or because I am a woman.

Here's my response to that: I was born before modern science could accurately determine the gender of unborn fetuses, but it didn't take scientific genius to determine my race. My mother was black. My father was black. That meant...I would be black. So, long before they knew whether I was a boy or a girl, they knew I was black. That, to me, says it all. I was born into second-class citizenry not because of my gender but because of my race.

While writing this chapter, I thought about another racial nuance that can be traced to the home. I love dogs.

From the time I was very young, my family always had at least one dog. My only regret about the peripatetic life I lead now is that it leaves me little time to care for myself, much less a dog. So I've never had a dog during the twenty-eight years I've lived in New York.

But I know dogs can sniff out those who are comfortable around them, and dogs can smell fear. Dogs don't see color.

Leslie Chambliss, a white colleague of mine, told me about the following incident and wondered whether it might be one of those little race things I was writing about.

"I was walking my dog in Greenwich Village one day, and when we passed by this black man, for some reason, my dog started to bark at him. I told her to stop barking at the young man. The guy came over to me and said, 'I bet you taught that dog to go after black people.' He was smiling, so I smiled back and told him that I did no such thing. And he said, 'Yes, you did. You taught him not to like black people.'"

Leslie wanted to know if I'd ever heard that kind of thing.

In fact, I had. In the 1950s and 1960s many blacks felt that whites had trained their dogs to attack black people on sight. Seeing Bull Connor using German shepherds against black demonstrators in Birmingham only fueled our fears. To this day there are blacks I know who refuse to buy that breed of dog. In days of segregation, blacks stuck to their side of town. Unless we were going to work inside white homes, we didn't drive through exclusive white neighborhoods, out of fear of being arrested as potential thieves or robbers.

Beware of Dog signs were visible warnings to us that "niggers would be attacked on sight." It was believed dogs raised in white communities were accustomed to seeing white faces, not black. Blacks were strangers. But the dogs smelled our fear . . . fear that came from being a black person suspected of trespassing on white property. The dogs responded to that fear.

That was then. Leslie was talking about now.

One of my black friends said the problem with black folks is that we want white folks to love us. I begged to differ, then I thought about Leslie's dog story and what Dennis, my nephew-in-law, had said: *White people treat their pets better than us.*

That's how some of us feel, and sometimes rightfully so.

In February 1999, Amadou Diallo, a twenty-two-year-old black man from West Africa who was standing, unarmed in the vestibule of his Bronx apartment building, was shot and killed by four white undercover cops. The officers said they believed Diallo was pulling a gun out of his pocket and was going to shoot them. They fired forty-one shots, nineteen of which struck Diallo, who was actually reaching for his wallet.

Black New Yorkers took to the streets to demand justice. They called upon fellow New Yorkers to join them in protests outside city hall to add their voices to the chorus of outrage. The silence from the white community was almost deafening. Even some of the white clergy remained silent about the Diallo case.

Around the same time this was taking place, a 431-pound tiger was discovered wandering the streets of a New Jersey suburb. Terrified residents contacted the police and the animal-rescue league. The tiger was finally located by rescuers with dart guns loaded with tranquilizers to subdue the animal. With television cameras rolling, New Yorkers—and, later, the rest of America—watched as the tiger was pursued and shot with the tranquilizers. When that failed to stop the tiger—and fearing he was headed toward a nearby residential community—the police shot and killed the tiger.

Well, you would have thought they'd killed Tony the Tiger. People questioned whether deadly force was necessary—this was, after all, a poor dumb animal. The woman who owned the tiger had others on her property, along with dozens of dogs and cats, and people called to find out what would happen to these animals and whether homes could be found for the dogs and cats, if not the tigers....

An unarmed black man is shot at forty-one times and blacks are left to seek justice in his behalf and defend his honor. A tiger is shot and whites question whether deadly force was necessary.

Don't Want a Nigger to Have Nothin'

During the 1950s the older blacks in my neighborhood used to tell this joke: A black man is seated in front of a white man on a bus. The white man notices a roach crawling on the black man's shoulder and removes it. The white man then tells the black man, "I saw a roach crawling on your shoulder and I removed it," to which the black man responds, "Put it back! White man don't want a nigger to have nothin'!"

Like my mother used to say: "Many a true word is spoken through a joke."

Blacks tend to become extremely agitated when whites appropriate something blacks consider theirs—through either heritage or custom, or by default. Whether it's our physical features—such as the way we walk, the way we talk, our style of dress—or it's our men or our women.

Edward Cooper of Washington said he took offense when some of the white students at the integrated university he attended wanted to join his black fraternity, Alpha Phi Alpha.

"We started Alpha Phi Alpha because we weren't allowed to join white fraternities at school," Edward explained. "So when this white male student came up to me and said he should be able to join any campus fraternity he chose to, I was upset. How dare he!"

"And we all know that once whites get in, they want to take over," added my niece Antoinette. "They want to be in charge, whether it's our fraternities, sororities, or our churches."

They want to adopt our children, although black families aren't allowed to adopt white children.

Yes, there is evidence for the statement contained in last paragraph. In 1992, the National Association of Black Social Workers, concerned about a slight rise in the number of white families adopting black children, issued a report which said that placing black children in white homes was a "form of racial and cultural genocide." Also, as recently as two years ago in 1998, there was a bitter court battle in Rhode Island between white foster parents and black relatives over a black child. The battle fueled considerable debate in the media and beyond.

Now they want our ghettos! In inner cities across the country, whites are reclaiming sections once considered undesirable if for no other reason than that "blacks lived there."

I was raised in a ghetto—a section of a city to which blacks, not Jews, were restricted—but I never thought of it that way. Not my little tree-lined block with its well-tended row houses. Ghettos were dirty, rat-infested slums, populated by people who didn't care about their community. That wasn't T Street in northwest Washington in the 1950s. There families planted grass, swept the sidewalks, and shooed children away from newly planted trees. There an entire block helped raise the children.

White folks could say what they wanted, little Lena knew a ghetto when she saw one and her neighborhood didn't fit the definition.

Not far from where I lived, along Seventh Street in southwest Washington, was a housing project. Located near the navy yard military complex along the Potomac River, the project overlooked the wharf. Within walking distance were several major federal government agencies. Talk about location....

The only time whites were seen in that part of town was when they were on their way to or from work, to shop at the navy commissary, or to browse along the wharf for fresh fish. Whites viewed the blacks living there from a distance—usually from inside their cars.

The blacks who lived there had tried for years to get federal officials to clean up the area and provide decent housing for the tenants, either there or somewhere else.

In the late 1960s the government launched an urban renewal project in the area. The old housing project was to be demolished. The tenants would be temporarily relocated to other public housing projects and allowed to return to the area once the new project was completed.

What was built there, however, wasn't exactly a "project." High-rise luxury buildings with terraces overlooking the water were constructed. Parks restored. Streets repaved. New streetlamps installed. Whites flocked to the area. The blacks who had lived there never returned.

Under the guise of urban renewal, blacks were moved out; whites were moved in. The pattern spawned a new term in the nation's vernacular: *urban removal.*

During the 1970s some blacks were fooled into the suburbs, thinking that life among whites, no matter where, was preferable to that among blacks. Postintegration census patterns show a tremendous exodus of blacks from inner-city communities to middle-class mixed-race suburbs. Their arrival prompted whites to move farther away—to so-called exurbs. As housing patterns continue to shift, whites are moving closer in to urban-renewed cities, where housing costs are now often beyond the reach of blacks.

While I was in Houston, Ria Griffin, a former reporter now working as a media consultant, took me to a black section in downtown known as Freetown.

It was so named because freed slaves settled there after emancipation. The streets are so narrow, residents park their cars along the sides of houses and dirt roads to allow traffic into the area. The frame shotgun houses that make up the neighborhood exude a kind of historic charm. Too fragile to support air-conditioning, many of the nineteenth-century homes are in desperate need of repair. But for generations of black Houston residents, Freetown represents a community of survivors. Whites who long ago deserted Houston's downtown area for the suburbs have seldom ventured into Freetown and wouldn't think they've missed anything in life if they never did.

Now, with a new baseball stadium located in downtown

Houston and a major urban revitalization program underway that has helped raise property values in the area, Freetown is being demolished to make room for new condominiums selling in the $250,000 range.

By destroying our communities, whites are taking away a vital part of black history. Those whites who want the world to believe that blacks live in squalor, that we do not tend our lawns or throw out our trash, that the arrival of a single black family into a white community automatically lowers property values, didn't grow up with the blacks in my old neighborhood.

There people swept not only their front porches but also the sidewalks. There kids received punishment for walking across a neighbor's grass or swinging on the limbs of a newly planted tree. In my neck of the woods, you didn't throw trash on the ground and just leave it there. You found the nearest trash container or faced a tongue-lashing for your oversight.

While I was working on this chapter, a story appeared on the front page of the *New York Times* about the destruction of row houses in Baltimore. The row houses with their characteristic white marble steps in west Baltimore were being torn down by the city. Historic preservationists were fighting to keep some of the structures standing for the sake of posterity. In the piece, a black woman recalled the days when residents would get on their hands and knees to wash the marble steps of their homes. I remember when I was a child hearing black Washingtonians marvel at the pearly white steps of the row houses and speak with pride of how the people of Baltimore took pride in the upkeep of their homes.

By destroying those houses, we destroy the memory—

recollections challenged by the passing of time. Nowadays the view most whites get of Baltimore is seen from Amtrak trains, and believe me, it's not a pretty picture. Gone is the Baltimore of my childhood, replaced by a disaffected black community, a shadow of its once glorious self.

"Why don't you live in Harlem?" asked a white woman who attended the focus group in Washington.

Her question was in response to complaints voiced by me and other blacks in the room that white Americans do not want to live around black people, period. The woman was suggesting that some blacks don't want to live around blacks.

The room grew quiet, anticipating my response.

Where I live on Manhattan's Upper West Side, a predominantly white, liberal-to-moderate section of the city, it's an eclectic community overflowing with amenities: two- and three-star restaurants, movie theaters, trendy boutiques, and a Starbucks on nearly every corner. Cabdrivers glance twice over their shoulders when I call out my destination. I like my neighborhood because it's convenient to public transportation, shopping, and a thirty-minute brisk walk from my office. That's why I chose to live there. Harlem had entered my mind.

Harlem is known throughout the world as New York's black residential and business hub. Unlike its image, Harlem is not a rat-infested, crime-ridden community. The people of Harlem are warm and friendly. Whites who venture north of 110th Street do not have to fear for their lives.

A renaissance took place in Harlem during the 1930s and another is quietly brewing. I shop in Harlem at the

Studio Museum and the 125 Mart. I socialize in Harlem and often attend church services there. I don't live in Harlem, because it's not convenient for me to do so.

There are few major supermarkets in Harlem. Those that are there, aren't conveniently located. There is only one cineplex theater, although another is scheduled to open soon. Delivery service is virtually nonexistent. Harlem residents complained long and loud enough, and finally a Domino's Pizza franchise opened on West 125th Street. Until Blockbuster Video opened a store four years ago, folks in Harlem either rented videos at outlets near their offices or at the few mom-and-pop stores in the neighborhood that stocked them. Harlem is conveniently serviced by the city's mass-transit system but not by "medallion" cabs.

There is also a housing problem in Harlem. Beyond the historic landmark section of Sugar Hill, with its magnificent brownstones, most of Harlem's housing stock is in a perpetual state of disrepair. This is not the fault of Harlem or its residents. Harlem is still suffering from benign neglect, which some blacks say is intentional. White developers have played to the stereotypes and refused to invest in Harlem and its residents.

The white woman who asked the question at my forum, and others like her, criticize middle-class blacks like me for turning our backs on the Harlems of America, for hiding behind: "I'd love to, but it's not convenient."

They get no argument from me there. But why should blacks be expected to live in black communities to prove... well, what? That we're black enough for white folks? Why aren't more middle-class whites living in Appalachia? Trust me, folks; it isn't the lack of basic amenities.

4. Little Things in the Workplace

The *New York Times* gives out monthly monetary publisher's awards to reporters, photographers, copy editors, graphic artists, and other editorial staff members for outstanding and noteworthy news and feature articles, headlines, photographs, and research.

When I arrived at the *Times,* in 1974, there were no more than twenty black reporters on staff and a handful of black photographers—most of whom were hired during the turbulent 1960s, when race was dominating the news pages. As in other news organizations, blacks had been hired primarily to cover the unrest—riots—in the black community. The most prestigious beats—White House, Capitol Hill, city hall, health and education, and foreign news coverage—remained the domain of white reporters at the *Times.* Because the publisher's awards are most often given to reporters covering major national news stories, black staffers seldom found themselves up for awards at the end of the month.

Although our work was often applauded by the *Times'* editors, our reports from the nation's inner cities paled in comparison to Watergate and the Pentagon Papers.

Al Harvin, the only black reporter in the *Times* sports department at the time, decided to implement his own awards. The C.O.O.N. Awards, chosen by Mr. Harvin, were given to black staff members for dispatches on the black community. The awards consisted of a dollar bill folded in the shape of a ring, which could be easily slipped on one's ring finger; a toy racoon; and an ink pen engraved with THE SISTER AND BROTHERHOOD OF SLEEPING-CAR REPORTERS.

Although our monetary compensation of $1 was $499 less than the average Publisher's Award, we accepted our C.O.O.N. Awards with zeal—congratulations were accompanied by squealing, high fives, kisses all around. Soon word spread through the newsroom that there were awards being given out only to black staffers. A few of our white colleagues asked about the rumor, but we didn't let on.

One day a white reporter was sitting next to a black female reporter when Mr. Harvin arrived at her desk and handed her a C.O.O.N. The reporter shouted to the blacks within earshot, "You guys, I got a C.O.O.N."

By now the white reporter's curiosity had gotten the better of him. He demanded to know more about these so-called C.O.O.N. Awards. Who, for example, decided which stories or photographs in the paper deserved a C.O.O.N. Award, and why were they only given out to blacks?

The black reporter told him the history, much the way I've presented it here, then she asked him, "Do you know what C.O.O.N. stands for? Committee of One Nigger."

When Mr. Harvin retired, in 1994, I told that story at his retirement toast. We all—blacks and whites—shared in the champagne and the laughter. At the time, though, it wasn't so funny.

Back then we felt slighted. Sure we cherished our C.O.O.N. Awards. I still keep one of Mr. Harvin's toy raccoons and ink pens on my desk at the office. But we also wanted to have our work recognized by the *Times* top editors and publisher. Because the system was set up in favor of the foreign and national beats and we blacks had yet to break the glass ceiling that separated us from those beats, our work, no matter how good, was being ignored. So we come up with a creative way to soothe our souls and what happens? Our white colleagues want a piece of that, too!

"Don't want a nigger to have nothin'."

What's Up with That?

Several years ago a black female reporter at the *New York Times* got into a verbal altercation with a white male editor there. The editor had removed the reporter's name from a story she had written on child prostitutes in Times Square. The editor said the reporter did not deserve to be included in the byline on the piece because it had to be rewritten by a white colleague. The reporter accused the editor of bias, saying that if the same standard had been applied to his work, he would not have won a Pulitzer Prize for his dispatches from overseas. The two argued for all in the main newsroom to see.

From my seat in the back of the newsroom, I noted the reporter was becoming increasingly agitated. With one hand on her hip, she presented her argument—to no avail. When she put her other hand on her hip, all of us—us blacks, anyway—knew all hell was about to break loose. Before we could intervene, a black male photographer and a black male reporter appeared from out of nowhere and gently pulled the reporter away.

"Let it go," they advised the editor.

Later that evening, the two black males were sharing a drink with the white editor, at Sardi's Restaurant.

"Let me tell you something about black women," the reporter told his white colleague. "When they're arguing and they have one hand on their hip, they're very upset. But when they put both hands on their hips, they're ready to kick ass. Remember that."

The female reporter admitted afterward that she thought she had actually hit the editor, an action that surely would have lead to her dismissal.

Years later the white editor said he saw a black woman arguing with a man and the woman had both hands on her hips. "She's very angry!" the editor recalled telling a white friend with him at the time.

Every day, blacks and whites work side by side, share coffee in the office cafeteria and, sometimes, a drink at the office hangout. Every night, they go their separate ways, not knowing any more about each other than they did the day before.

The workplace has become a battleground of racial unrest. Job discrimination suits are as commonplace today as they were in the 1960s when the Civil Rights Acts were passed. Blacks believe that whites are threatened by the presence of blacks in highly skilled, good-paying jobs.

Whites often say they have been bypassed for promotions, because of affirmative action. Despite the best diversity initiatives, there is still tension in the office.

Blacks hear "diversity" and think: *Sound and fury signifying nothing.* Whites hear "diversity" and think: *White managers are being forced to bend over backward to accommodate minority-group members.* Neither group feels anything will come of these initiatives. They are, say blacks, a way for the white man to ease his conscience and feel good about himself. They are, say whites, nothing more than sets of recommendations that will soon outlive their usefulness as effective workplace tools.

Whenever the issue of diversifying the workplace comes up, whites will inevitably talk about hiring "the most qualified applicant" or promoting "the most qualified person." Implicit in what they're saying is that if you're black and you get the job or the promotion, it's not because you were qualified but because you were black.

"Whites act as though they all got their jobs because they were qualified," said Mr. Mitchell, a black man who participated in the Washington focus group. "For years whites got jobs because their dad played golf with the boss or their mother belonged to the same bridge club as the boss's wife. Never mind that they may not know a damn thing about doing their job; they got it. But the minute we are promoted, they want to talk about racial quotas or affirmative action."

Pauline Schneider, who works for a major Washington law firm and shared her thoughts during the session in Betterton, Maryland, noted that when her firm was in the process of hiring summer interns, one of her partners addressing the issue of increasing the number of minority summer hires always and only used the word "qualified" when speaking of minority candidates.

"I took exception," said Ms. Schneider. "I wanted to know why he felt it was necessary to mention 'qualified' each time he referred to a minority candidate when everyone we hired to work at the firm was expected to be qualified to do the job."

During the 1990s "diversity" became the mantra of corporate America. A cottage industry evolved to handle requests from corporate executives concerned that the workforce was changing and the workplace would have to change with it or face extinction.

According to the Workforce 2000 report, prepared by the U.S. Census Bureau, by the year 2000 the workforce—once a predominantly white male domain—would be largely minority and female. Employers who moved forward to diversify their workforce and find ways to respond to the needs of this workforce would be ahead of the competition and establish a foothold in the twenty-first century. In response several companies embarked on efforts to diversify their staffs.

Thomas Kochman, a diversity consultant from Chicago, Illinois, has counseled dozens of Fortune 500 companies and city governments on how to handle a diverse workforce. He has also written extensively on the subject. Over the years, he's developed his own way of doing things, a style that does not always sit well with white managers.

Mr. Kochman's workshops ask participants to accept a few "self-evident truths," chief among them that all is not equal in the workplace. He also talks about archetypes, which he defines as widely accepted "truths" about groups, that, unlike stereotypes, are generated from within the group. The idea that women tend to talk about their personal feelings and that men don't is an archetype.

Among the archetypal Anglo-American behavioral traits he lists for males are aggressiveness, independence, dominance, self-confidence, sexual assertiveness, pragmatism. For females he lists: passivity, dependence, nurturance, feelings of inadequacy, sexual receptiveness, and conformity. He always notes that there are exceptions to the rule and that these traits apply to most but not all persons in a particular race group. Among the African-American behavioral traits he lists for both males and females are aggressiveness, independence, nurturance, self-confidence, sexual assertiveness, emotional expressiveness, nonconformist attitudes.

In describing differences between blacks and whites in terms of their styles of work and play, Mr. Kochman notes that blacks tend to be more spontaneous and improvisational in their styles. They also tend to be more expressive in nature and to personalize tasks. Whites, on the other hand, tend to be more methodical and systematic in their approaches to work and to be more restrained. They are also more likely to view their daily work as "role oriented."

"In communicating with others, blacks' cultural orientation is to 'do what you say,'" said Mr. Kochman. "In other words, 'walk the talk.' The focus is on personal-action results. Another way of putting that is 'If you're not part of the solution, you're part of the problem,' is how blacks see it. Whites, however, expect people 'to mean what you say,' and the focus is on intent. Or 'you get out what you put in.'"

Such cultural differences can have a profound impact on how blacks and whites go about their jobs. Many blacks would rather have a personal face-to-face confrontation with the boss to hash out problems than go about their workday acting as if all is well. In other words, tell it like it is. Kochman has found—and here I tend to agree with his findings—that whites prefer peace before truth. "Better to have an insincere peace than a sincere quarrel," Kochman noted.

I, for one, do not like written critiques or evaluations. Why? Because there are certain things that make their way into written correspondence that would never be said to my face. A white male editor once wrote the following query on a story I'd written: ***"Lena, What Does This Mean????"***

Now, think of how that query might sound spoken. Get my drift?

I mentioned this to him and he immediately apologized. Apparently two of my white female colleagues had also approached this editor to complain about the tone of his written inquiries.

Also, whenever I hear that someone is a "team player," I interpret it as white speak for someone who plays by the rules—which, as many blacks know, were created by whites. Don't get me wrong. I've been called one of the best team players around. I've played along to get along. But I sometimes feel like that old Muppets cartoon I have clipped to my bulletin board: "In our office, we're team players. Now if we can only figure out which game we're playing."

If whites wonder why blacks are often reluctant to subscribe to the team concept, it's simple: They—*we*—are afraid we won't get credit for contributions made, no matter how large or small. And too often blacks find themselves being asked to put the team's considerations first, and that means putting white preference over black.

"We sacrifice for the team and whites get all the glory," said Quinn Walker, a young black man who lives and works in Washington. "If all things were equal, I'd be more willing to buy into the team concept. But the playing field isn't level. And I doubt it ever will be. I think whites know blacks are reluctant to subscribe to the team concept. But when we say so, we're accused of being stubborn or uncooperative."

Going along to get along never worked for blacks. We feel we're damned if we do, damned if we don't.

Whites, on the other hand, resent the fact that many companies are bending over backward to change corporate culture that has been a proven success for the companies and their white employees.

Blacks (add here Hispanics and women) haven't made these Fortune 500 companies into Fortune 500 companies. That credit goes to white men. (No argument from me there, tough as it may be to admit.) White employers feel they have not been given credit for what they have already done to make things equal in the workplace. Indeed, blacks' median incomes are almost comparable to whites', and more blacks than ever are in positions of "power." Why not focus on what has been done, rather than what hasn't? whites have argued. But even more to the point, when it comes to the workplace, many employers are "equal opportunity oppressors."

Susan Sherwood, a white postal worker based in Bethesda, Maryland, recalled trying to comfort a black fellow worker who'd had a run-in with his white supervisor.

"This particular supervisor ordered the guy into work, even though he said he was sick," Sue recounted. "He came to work, and the two of them really got into it. So I told my colleague that the supervisor is a prick and that he did that to me once. Well, my black colleague got furious. 'Sue, this may have happened to you once; he does it to me and the other blacks around here all the time.'

"He probably was right," Sue went on. "But I said that to make the point that the supervisor was a prick and that my coworker shouldn't let him get to him."

The workplace is also filled with examples where race is perception but not reality.

Dr. McFarland, who works in pediatrics at Long Beach Memorial Hospital, in California, shared this story with me about how the fears and rejection she has experienced as a black person have influenced the way she behaves and thinks about other people. "I was new to the staff. The hospital offered free lunch to physicians, so I started going to the cafeteria to eat. I would look around the room for someone to sit next to. And because I was new and didn't know anybody or recognize anyone, I would usually end up sitting by myself. And I'd sit there thinking about how I hadn't met anybody and how the system was so prejudiced in that no one had introduced themselves to me. So I decided to analyze what was happening. Once before, I'd taken the initiative to go up to someone and introduce myself—which is usually what I do; I'd always felt comfortable introducing myself to other black professionals—the one person I'd made acquaintance with at the hospital was a young Hispanic woman. I wondered why. Because she was Hispanic, maybe. I'd liked the way she dressed, but why had I singled her out?

"Then I realized," Dr. McFarland went on. "Most of the other doctors were white males who were older than me. I began thinking they wouldn't accept me because I'm a black woman and I have dreadlocks and I don't look like your typical pediatrician. I was getting all worked up. Then I said to myself that the same thing I think they're going to do to me, I was doing to them—making assumptions based on outward appearances. Finally, after six months of this psychoanalysis, I went up to a table where a group of white surgeons were sitting and asked if I could join them. They pulled out a chair for me."

What struck Dr. McFarland about the experience was that for six months, five days a week, she had allowed the past to affect the present.

"So now I see how difficult it is, not only for whites to judge me and interact with me as a black woman, but also for me to be free and accepting of them."

An ABB—All Blacks Bulletin

Yes, it's a small world. But, for the life of me, I can't understand why so many whites seem to think all black folks know each other.

During segregation, blacks were forced to stay—and stick—together. Today, although still largely concentrated in major urban areas, blacks are dispersed across class and region. Nevertheless, we are placed in the uncomfortable position of being expected to know this or that person, be they politicians, rap artists, entrepreneurs, or Jamals-on-the-street, simply because they're black.

One of my former colleagues at the *Times,* a black woman, told me once that a white female editor had asked her if she knew Biggie Smalls, the heavyset black "gangsta" rapper who was fatally shot in 1997. My colleague had been raised in an upper-class home in Los Angeles. Her family vacationed in Lake Tahoe. She moved and shook in exclusive Hollywood circles, places where you'd find Quincy Jones and John Singleton, not Tupac Shakur and Biggie Smalls.

"Yet, I'm being called at home and asked if I know Biggie Smalls or had friends who had contact with him," my colleague recalled.

For years black journalists have complained that their news organizations tend to rely on them to provide an edge in coverage on "black stories," while ignoring them as possible sources on stories involving white Americans.

When a federal office building in Oklahoma was bombed in 1995, white editors, for the most part, didn't seek out the black journalists in the newsroom. We weren't expected to know anyone in Oklahoma City, a predominantly white town in the Midwest. We couldn't possibly know members of the subversive white group suspected in the bombing. And, for sure, the best Justice Department sources couldn't possibly be contained in the Rolodex of a black reporter.

But in the spring of 1992, when riots erupted in South Central Los Angeles after four white police officers were acquitted of violating the civil rights of Rodney King, a black man, black journalists' phones were ringing off the hook.

An "all-blacks bulletin" was posted in most major newsrooms, my own included. Calling all blacks, calling all blacks. We need contacts, sources, black bodies to "send into the ghetto."

In no way am I suggesting that white editors not seek out black reporters for help or insight into such news stories. But why not also turn to these reporters for stories involving white Middle America?

It's not just newsrooms where this complaint is heard. Blacks who work in other areas of the private and public sectors say their opinions are seldom sought by white managers on issues like foreign affairs, the economy, or the environment. But when the discussion turns to subjects like poverty, crime, or single parenting, blacks are deemed in-house experts.

"My boss, a white man, once asked me how food stamps work," said Alisha M., a black administrative assistant at a New York hospital. Ms. M. asked for anonymity because she did not want to alienate her boss. "Neither I nor my family has ever used food stamps, so I wouldn't know. I told him that and then asked him why he asked me. He said he thought I might know, that's all."

Ms. M., a woman in her late fifties and the only black in an office of five, said she has been asked to speak on black subjects ranging from the music of Lauryn Hill, the young black Grammy Award–winning hip-hop artist—even though she didn't know who Ms. Hill was until her grandson told her—to the state of Jesse Jackson's marriage. She was even asked whether she knew anything about "wilding." The term became widely known during the early 1990s after a white female jogger was raped and severely beaten in Central Park by a group of black youths who were roaming through the park intimidating anyone they happened upon.

"I play the stock market, I go to church, I vote in every election, but I'm not the first person asked about investing or religion or politics," Ms. M. noted. "No, I'm not supposed to know about such matters. But food stamps and wilding... I'm the expert."

Twice As Hard, Half As Far

The belief that blacks have to work twice as hard to get half as far has become an accepted truth among workaday black Americans, be they judge or janitor.

"Let a black person had done that" is black dialect for "Whites can get away with bad behavior like incompetence and infidelity in the workplace. But if black people so much as try those things, they're fired."

Knowing that our every move is likely to be challenged has made us stronger in some ways, if not a bit compulsive in our behavior.

Freeman Moyler, an imposing six-foot-three-inch light-skinned black man with white hair, managed to rise from the position of elevator operator at the *New York Times* to Vice President of Building Services. When asked by a younger generation of blacks at the newspaper how he managed to turn a deaf ear to racial slights directed his way, Moyler, who died in 1998, told this story: "One day after I started working here, I was running the elevator and two white men got on and told a 'nigger joke,'" Mr. Moyler said. "That night, I went home and when I was in bed with my wife, I asked her to tickle my feet with a feather. She did but wanted to know why. I told her that I wanted her to do that every night until I learned not to laugh at the sensation. When I learned that, I learned how to turn a deaf ear to nigger jokes."

Race-based idiosyncrasies are commonplace.

There are black professionals I know and have worked

with who routinely tape conversations with white managers or bosses because they fear their word will not carry as much weight as that of a white colleague and they may one day need solid evidence to absolve themselves of wrongdoing or to prove themselves right. There are black journalists, list me among them, who often overreport stories because they believe their news accounts may be subjected to greater scrutiny. I've heard of black lawyers at prestigious firms who refuse to meet alone with white clients, concerned their reputations could be ruined by saying the wrong thing to a client distraught over legal problems.

"It's automatically assumed that whites already know the work or know how to do the job," said my nephew, Frank Grinage Jr. "We aren't supposed to know. We automatically have to be shown the right way to do something."

"I think whites aren't conscious of the fact that they might be doing something that's racist," Ronald Frazier said in response to Frank's observation. The two attended the same focus group at the Hyatt Hotel in Bethesda, Maryland. "For example, we were at a meeting and a good idea was put on the table by an African-American woman. It went around and finally we couldn't reach a decision and it seemed that the best idea was the first one, the one from the black woman. The moderator, who was white, said, 'You know that brilliant idea that came from Dr. So-and-So, that's exactly what we need to go with. But it wasn't Dr. So-and-So, who was also white, it was the black woman.

"The moderator assumed that the idea had to come from the white man and not the black woman," continued Mr. Frazier. "It may be that he forgot who put forth the first idea, but instinctively and without second thought he attributed it to the nonblack. That's where the racism comes in. That's the unfinished business Clinton was talking about. I've sat on panels where the black was clearly the most qualified, only to have some white say, 'You know, there's just something about him or her that I'm not completely comfortable with. My gut says we should go with Bob. Bob would fit in better.' "

Judge U. W. Clemon, a federal district court judge in Birmingham, Alabama, says he is constantly being asked to recuse himself from cases, because of either his background as a civil rights activist or his close ties to Birmingham's former mayor Richard Arrington, who like Judge Clemon, is black. At the same time, he noted, his white colleagues on the bench—some of whom were staunch segregationists or have close relatives who were affiliated with the Ku Klux Klan—are seldom, if ever, asked to remove themselves from cases in which race may be a factor.

"Civil rights cases were basically handled in the federal court here, and a lot of people were concerned that I would not forget my experiences when I became a federal judge," Judge Clemon told me during an interiew at the Tutweiller Hotel in downtown Birmingham. "You strive for perfection and the idea of objectivity. Sometimes you don't make it. In the cases I hear, if I feel I can't be objective, I recuse myself. Because, at the end of the day, I'd like to be able to say I haven't done to anybody what I wouldn't want done to me.

"Lawyers are always asking me to recuse myself, usually for some facetious reason, but I'm always being challenged," he added. "I think I write more decisions denying motions to recuse than any other judge in our circuit; maybe than any other judge in the nation."

According to Judge Clemon, shortly after he was confirmed to the court by the U.S. Senate, a white Democrat who sits on the court of appeals and who had actively tried to block Judge Clemon's confirmation, let it be known to Judge Clemon that "every time one of my decisions came up before the court of appeals, he was going to reverse it."

"And he did on a number of cases," Judge Clemon recalled, managing to laugh at the memory. "This was the same man who was president of the student body at the University of Alabama, when UAB was resisting desegregation."

Knowing his every decision was being watched made Judge Clemon feel like "every *t*, had to be crossed."

"And so, yes, you do have to be extraordinarily careful, so that I know that there's no way they can reverse me on the merits of the case," he explained.

Says Ria Griffin, the Houston media consultant who showed me Freetown: "You've got to be smarter than they are, work harder than they do, and look better than them. Might as well be prepared for it."

When you do manage to climb the corporate ladder, be prepared, Ms. Griffin said, to have your achievements belittled by whites as "an affirmative action."

Added Alvin Wright, a black Houstonian working in the field of communications who attended the forum in Houston with Ms. Griffin: "Whites are still getting the best jobs, the top salaries, the big promotions, yet they begrudge you [blacks] if you get so much as an award for doing a good job. It's not as if blacks have come in and taken over these Fortune 500 companies or local government. We haven't, 'cause they won't let us."

Other black professionals say their perceived status—the way they dress, the cars they drive, the children they have in private school—is a cause of curiosity, some say envy, to many whites.

"How can you afford so many different clothes?" my sister-in-law Clarice Williams has been asked by some of her white colleagues. "You wear a different outfit every day."

She doesn't; it only looks that way to a people who—for the most part—never had to learn how to stretch a dollar or turn scraps from the table into ethnic cuisine. Blacks like to think they know how to survive. Our ancestors had to make a little bit go a long way. They managed to raise families and, in some instances, put kids through college on the quarters and dollars they earned scrubbing floors and operating elevators.

So when whites see blacks doing well, they may be under the mistaken impression that we're well off. There's a difference.

"In my office I'm the only one there raising three kids, [who] has two car notes, a house note, and is still better off than nearly all of them," said Don Payne, the Houston police department's communications director. "But I bring my lunch in every day, and the reason I do is because I've got to try harder to make more out of the little bit I've got than they do. And I've gotten good at it because I've been doing it for a long time."

Many blacks will tell you, we've also learned to measure our worth not by standards set by a Protestant work ethic, for surely then we'd never measure up, but by historical evidence that proves us to be a proud, hardworking people.

"Don't expect nothin' from white folks on this earth," the old folks often told a younger generation of blacks foolish enough to believe otherwise. "That way, you'll never be disappointed."

A white female colleague once said to me that she felt blacks at the newspaper and at the public relations office she'd previously worked at, seemed to take disappointment—whether being turned down for a promotion or a salary increase, or receiving a negative evaluation—in stride.

That may be due to the fact that a people deemed less intelligent and less worthy than white Americans could not possibly expect to amount to anything in white America's eyes. Now that may seem like a laissez-faire attitude, but I believe it's helped us survive as a minority in a majority world. I've seen white classmates and colleagues literally contemplate suicide because they received a failing grade on a report card or were turned down for a prestigious assignment.

"Just 'cause they say you're shit, doesn't mean it's so," I said to a white friend who was suffering considerable angst after receiving a negative evaluation.

In 1998 I was elected Chairperson of the *New York Times* Unit of Local #3 of the Newpaper Guild, becoming the third black to hold that position at the paper, and the second woman.

I didn't exactly seek my position. I was talked into it by a smooth-talking Irish colleague, Tom Keenan, who had the audacity to play a race card on me by suggesting that the interests of black workers at the company might be better served if more blacks were involved in the guild.

He had a point. Blacks have always felt the union movement was antiblack. As recently as the mid-1980s, many of the trade unions refused to allow blacks in, barring their path through Byzantine restrictive clauses that required union work be performed by card-carrying union members. The only way to become a member was to get the work. Without work, you couldn't become a member. A working man's catch-22.

(Spike Lee, the black filmmaker, has managed to get several blacks into the film industry trade unions by giving them work on his movies.)

When Tom asked me to join his slate at the *Times,* I agreed, accepting his assurance that it wouldn't require much of my time.

"Maybe, a few hours each month," he said, trying to assuage my reservations.

Should have known not to trust a white man offering something for nothing. (I'm smiling here, OK?)

In 1997, when Mr. Keenan became ill, I assumed the role of acting chair, a position that often required my looking beyond race, gender, and the proverbial "right or wrong" to champion the union's cause.

It hasn't always been easy.

On onc occasion I found myself in an enclosed office, trying to persuade a group of white male managers that they shouldn't fire a white female employee for telling a racist joke overheard by a white coworker who found it offensive. On another, I defended the company's decision to suspend a white manager for transmitting a derogatory Ebonics message that offended several black employees who felt that anything short of firing would be unjust. I've turned deaf ears to black coworkers who injected race into their defenses as an excuse for dishonesty, lateness, or failing to do the job. And I've heard myself say, more than I care to admit, that "everything between blacks and whites isn't about race."

What's up with that?

When did Lena, the black-power militant, the black nationalist, the hostile young woman who didn't start speaking to whites until March 19, 1974 (the day I started working for the *Times*), turn into this detached, fair, objective, politically correct mediator of race and all its indignities?

I'd like to think that I've always been guided by a sense of fairness: You know, "Right don't wrong nobody." That's not something I learned in journalism school. I learned it at my mother's knee. Had it instilled in me in Sunday school service at Tenth Street Baptist Church, in Washington. It's been my creed and it's served me well.

I wasn't sure I'd be good at the chairperson job (maybe that's a black thing, this tendency to be riddled with self-doubt). But it's given me a window on corporate management that I might not otherwise have had and it's transformed me into what I jokingly call a white manager's worst nightmare: a black woman who thinks like a white man.

During our contract negotiations in 1994, a white male manager told the guild's representatives that seniority had outlived its usefulness in the workplace and management was examining ways to determine salary increases, bonuses, and promotions based on merit rather than seniority. That would be a "fairer" measure than seniority.

Now, I'm sitting there listening to this and wondering. Finally, I explained to him that his view of "fair" wasn't quite the same as mine.

"You know, when I was a little girl growing up, I

wanted to work for a big company like this, but my mother told me I couldn't because black people weren't allowed to do certain jobs at white companies," I said. "When I told her that wasn't fair, she said it wasn't, but one day things would change. And they did. Then, when my people began to work at these big white firms, we were often the last hired and the first fired. And we were told that's because there is a system of seniority. So the person who's worked at a company the longest shouldn't be the first to be laid off. Now, I didn't like that one bit, but it seemed fair to me. So my mother told me to 'find a good company to work for, and to put in your years, and you'll get your seniority status.'

"So now you are telling me," I went on, "that after people like me have finally reached a point where we can exercise our seniority status, management wants to abolish it. And that would be fair. Well, understand if I don't agree."

He told me my point was well-taken.

Let's just say this: Seniority still prevails at the *Times*.

The Conspiracy Theory

It never fails. Whenever two or more blacks get together at any one time in any one place, inevitably some white person thinks something's up.

"Is this a conspiracy?" a white male colleague asked me and two other black colleagues who were hovering around my desk one afternoon.

Ours had been an impromptu gathering. I do not recall what brought us together that particular day, but surely it was nothing of monumental consequence. Perhaps we were doing what we are sometimes prone to do: come together to share a friendly thought, a word of professional advice, a query of concern, or just to shoot the breeze. Even if we had come together to gripe and moan about work, life, or whitey, why was our white colleague so quick to assume our actions were conspiratorial in nature?

We responded to his question with a snide, collective "*Yes!*" and a look that said, "Get lost."

I'm sure, as were my black colleagues, that our white coworker interpreted our response as another example of black rage. In part, maybe, it was. Among Jim Crow's many lasting legacies was a rule that prohibited no more than three blacks to gather at any one time in public. The law was designed to keep blacks from conspiring to do harm to whites.

Whites—and some blacks—would like to think that the rule no longer applies. Legally, it does not. But in practice it is still alive and well.

In interviews with blacks, especially young black males, I was told of countless times groups of blacks, regardless of age or manner of dress and demeanor, had been approached by police officers and questioned about the purpose or nature of their gatherings.

"So what brings you all to this part of town?" Robert Simpson, a black college graduate and computer technician, recalled being asked by a white Washington, D.C., policeman. Robert and a group of his fraternity brothers had briefly paused on a street corner in Georgetown to figure out which of the area's trendy watering holes they might patronize. "We explained that we were going to have a drink. He looked at us with a mixture of contempt and suspicion, then said we shouldn't be hanging around on the corner. We again said we weren't loitering; we had merely stopped to gather our thoughts and bearings. He then told us to move along."

Would a group of twenty-something white males have been questioned? Who knows? But Simpson recalled that at the time he and his black friends were being hassled by the cops, a group of white youths, dressed à la grunge, were on the opposite street corner, laughing and joking and not being hassled by the police.

The idea that a congregation of blacks is engaged in a conspiracy speaks to the level of distrust that exists between blacks and whites. Blacks from all walks of life have, at one time or another, expressed the belief that it's really white America that is engaged in a conspiracy—one designed either to eliminate blacks or to adversely change their lives.

The belief has given rise to speculations that range from the ridiculous to the absurd and has prompted a kind of widespread paranoia that "they're out to get us." Which, by the way, doesn't mean they aren't.

History lends credence to the conspiracy theory, the one blacks believe whites have, and still are, waging against the black race.

The Tuskegee Syphilis Study, a forty-year government-sponsored project, wooed infected black men living in and around Tuskegee, Alabama, as test subjects, under the guise that they would be treated for the sexually transmitted disease, when, in reality, treatment, including penicillin, was withheld. The project ran from 1932 to 1972, during which time the 399 black men infected with the disease suffered the devastating ravages of its effect.

The Tuskegee study, along with Co-intelpro—a covert operation conducted by the FBI against subversive elements in America, including Dr. Martin Luther King Jr. and the Black Panther party—lends credence to a prevailing attitude that conspiracies against black Americans aren't all that far-fetched.

While it's true that Co-intelpro also targeted white groups like the Weathermen, there was never a plan to saturate poor white communities with drugs as a way of undermining the terrorist campaign being waged by the Weathermen. But there was an FBI plan to undermine the Panther movement by allowing drugs to flow freely into black sections of Oakland, California, the headquarters city of the Panther party.

Now don't get me wrong on that last point! I would be a hypocrite to blame J. Edgar Hoover or the FBI for the drug problem that has ravaged inner-city black communities, destroying families and entire neighborhoods in its wake. As the adage goes: You can lead a horse to water, but you can't make it drink. Blacks can't use the excuse that they brought the drugs into our communities and hooked us on them. No one forced drugs into our arms or up our noses. And the irony of it all is that the plot worked... better than Hoover himself might have imagined. Several members of the Panther party, including its charismatic leader, Huey Newton, along with two of his closest associates, Eldridge Cleaver and Bobby Seale, became drug addicts.

There are blacks who believe that AIDS is part of a genocidal conspiracy against them. The word on the street is this: The scientific community wanted to test HIV, the virus that causes AIDS, to determine its effect on animal life, but the virus got out of control and there was no way to stop it.

A *New York Times*/CBS-TV News poll in 1990 found that one black in ten believed the AIDS virus was "deliberately created in a laboratory in order to infect black people," and an additional two in ten thought that might be so. Even the various treatments for and preventives against AIDS are suspect.

Some blacks—Dick Gregory, the comedian, among them—even believe that HIV has something to do with diet and that blacks in particular are being targeted. As a result many blacks refuse to accept data that show the virus to be spreading at an alarming rate among blacks and often dismiss this information as "propaganda."

"It gives whites something else to use as a basis to discriminate against blacks," Mr. Gregory told me once, during an interview on the subject. "We have enough stigmas against us. Now they want to say we're more likely to contract AIDS than whites, even gay white men. Come on! Are we supposed to sit back and believe that crap? Now they're saying there is a higher incidence of cancer among blacks. They always quantify by saying, 'in proportion to their percentage of the population,' but that fact is always glanced over. It's a subtle form of genocide."

Many black Americans are reluctant to become organ donors, because they fear the white medical establishment will cannibalize critically ill or critically injured blacks to use their organs as transplants for white patients.

Research shows that blacks are disproportionately

overrepresented among people waiting for organs and underrepresented among people receiving transplants. And therein lies the rub!

Because of fear that blacks will not receive donated organs, fewer blacks donate. Many blacks say they distrust the white medical establishment and are convinced there is discrimination in selecting organ recipients. Blacks believe that if two people—one white, one black—needed an organ, doctors would give it to the white patient. Blacks also believe that if they were in a car accident or otherwise critically injured and had signed a donor card, doctors would declare them dead prematurely, to obtain organs for whites.

This seldom-discussed subject started to reverberate through the black community in December of 1967, shortly after Dr. Christian Barnard of Capetown, South Africa, became the first physician to successfully perform a human-heart transplant.

What many blacks remember about the surgery was a little-publicized fact: The donor of the heart was a black man and the recipient, a fifty-five-year-old white grocer.

At the time, South Africa's system of racial apartheid denied the country's twenty million blacks the right to vote, to own land, to work in certain jobs, and other basic freedoms that the nation's two million white Afrikaners took for granted. Black South Africans weren't good enough to shake white people's hands, but a black man's heart was good enough to place into the body of a white man.

Black folks in my neighborhood managed to put an interesting spin on the story.

"It took a black man's heart to make the experiment a success. How 'bout that," my grandfather John Adams told me. "We got real big hearts, us black folks. That's what that tells you."

My own family members are divided on the issue of donating organs. My sister Gloria has always felt that when her day comes, if her organs can be used to help someone else live a longer, healthier life, why not. My sister Ada and I have always jokingly replied that we'd prefer to leave here with what we came with.

And the Winner Is...

Another ordinary thing that riles many of us and fuels mistrust between the races is the assumption that we blacks are supposed to accept white people's word as the truth, the whole truth, and nothing but the truth, when our basic instincts tell us otherwise.

For example, blacks are expected to accept the Nielsen ratings as an accurate barometer of television-viewing tastes when I, and scores of other blacks, have never known or even heard of one black family that's had a Nielsen Box.

Now, that may seem like a little insignificant thing to white Americans, but such trifles become increasingly wearing and discouraging to blacks.

My newspaper colleagues and I rely heavily on data obtained through the *New York Times*/CBS-TV polling system. This telephone polling system, which was first used at the paper in the late 1970s, polls hundreds of Americans about various issues. The *Times* and CBS, like many other media institutions, have used the system to "accurately" gauge Americans' attitudes and opinions on a range of subjects. But blacks view such polls with a degree of skepticism.

Why? Again we're hard-pressed to find blacks—unless specifically targeted on a particular issue—who have been randomly selected to be polled.

Some experts suggest that black Americans, like many Hispanic Americans, are less likely to have telephones in the home. Others say that many minorities are less likely to be at home during the daytime hours when many of the polls are conducted. Still others suggest that minorities are often distrustful of taking part in surveys and polls conducted by white organizations, especially those conducted via telephone, because they are not sure how the information will be used. We know for a fact that census data is used to determine everything from state aid to congressional representation, and municipalities have exploited the figures to obtain more money and more votes in Congress.

Whatever the reasons, the widely held belief that blacks are omitted from such routine polling raises a degree of suspicion within segments of the black community as to their validity.

And another thing, I, like so many other blacks, have more than a few problems with the methodology.

Media groups poll anywhere from 900 to 1,400 people (a tiny percent of the 270 million–plus people in America), and from that sampling, which has a margin of error of plus or minus three percentage points, make quantum assumptions about what "all" Americans think.

"They can do that, yet question blacks when we say things like 'we' don't trust white people," said Edward Cooper of Washington. "If they can talk about how the majority of Americans feel about something based on questions with a few hundred people, surely we should be able to say what blacks feel based on our collective history and our conversations across America with each other."

Would You Like to Be My Mentor?

Several years ago, when I was still a young, struggling journalist, I confessed to one of the senior editors at my newspaper—a man who could make or break careers—that during the year I'd worked as his news clerk and the years up to when I was promoted to a position of staff reporter, he'd never invited me out to lunch, a courtesy routinely afforded several young white colleagues who'd taken the same climb up the ladder. I'd like to say such a mundane thing didn't matter to me, but it did, and I told him so.

"Do you know what message that sends when the staff sees you taking a young clerk or reporter out to lunch?" I asked him. "It says that person is special in your eyes. That you want to take them under your wing. For all we know, you may be taking the person out to read them the riot act, but we don't know that. We see this kid going out with you and then say he or she is going places."

The editor seemed sincerely surprised by my personal revelation. Within days he called me and we went to lunch at Orso's, one of the eternally "in" restaurants on West Forty-Sixth Street.

"You never confided in me like [so and so]," the editor told me, speaking of two white female reporters who previously served as his clerks. "They'd tell me when something was on their minds. [Such and such] loved to bake little cakes for me, and [so and so] said I was like a father to her."

Respectfully I explained that it was unlikely that I, a young black woman from Washington who'd lost her father at age thirteen, would see him, a Jewish man, as a father figure. As such, I probably wasn't as likely to tell him my inner thoughts and fears, like my two female colleagues who were both Jewish.

He seemed to understand, but what surprised me was that something like that had never occurred to him.

Godfathers, rabbis, fairy godmothers have become synonyms for *mentors,* the senior managers who take lower-level employees under their tutelage and help guide their careers upward. Seldom do white managers seek out blacks to mentor. We are often left searching for a sympathetic ear in a position of power, to ask for advice or guidance.

Regardless of race, people tend to choose protégés who remind them of themselves, in their attitude, background, or experience.

Many a career has been made by a skillful mentor who took another employee aside and told him or her "how things are done around here," and the federal government couldn't do a damn thing about it.

Black professionals I spoke with told me about young white colleagues who had mentors do the work for them. Others recalled white mentors who made sure their white protégé sat next to the chairman of the company at lunch or dinner, and white mentors who introduced that "white kid who was going places" to all the right people.

They'd tell them how to dress, how to talk, which job assignments to take and which ones they should turn down. They pulled them up the ladder, rung by rung.

There are exceptions to the rule. I've had more than

my share of mentors, whites and blacks in virtually equal numbers. They whispered in my ear when they felt I was going astray, convinced me to do things I once felt incapable of doing, and guided me along my merry way.

Unfortunately my experience does not reflect the rule. Too often men mentor other men; whites other whites; blacks other blacks. And with white males still holding the most prestigious jobs, still heading all but two Fortune 500 companies, they are in a better position to assure that the line of succession reflects the status quo.

Whites argue that blacks are as guilty of choosing their own kind to mentor as whites, and they are right. Whites are also well aware of the countless black professionals who, having reached the top, refuse to pull another black along, or who choose a white protégé. Several companies, my own included, have tried to broaden and diversify the pool of mentors by implementing mentor programs in the workplace. Sometimes they work; sometimes they don't. Mentor-protégé relationships work better when they are formed naturally.

Blacks still do not trust whites and may be reluctant to open up to an assigned white mentor, no matter how well-intentioned. Whites may be fearful of doing or saying the wrong things around a black protégé and thus be reluctant to serve as his or her mentor.

Yet it is exactly these kinds of relationships that can help foster understanding across racial lines.

The little things in the workplace appear to aggravate blacks and whites more than the little things that transpire outside our work environment. Maybe because we feel we have so little control over what happens on the job. If a white person doesn't want to live next door to a black person, she doesn't have to. She can purchase a home in a predominantly white part of town and not be called racist for doing so. If a black person doesn't feel comfortable going to a predominantly white church, he can easily find a black church to fulfill his spiritual needs.

We can choose our neighborhoods, our schools, our churches. We often can't choose our jobs. The job chooses us. And if the money is right, trust me, no one other than a fool is going to turn it down because "too many blacks work there." There are isolated cases, but they are the exceptions rather than the rule.

When schools in several of our major cities became "too black and Hispanic," whites simply transferred their kids to private schools. But when your company becomes more diverse, it's not always that easy to simply transfer to another job. Few of us want to piss away seniority and a pension simply because we "don't want to work around certain folks." So we are forced to tolerate one another for a minimum of eight hours a day.

5. Little Things in Social Settings

Is This a Party or What?

SHELBY Steele, a black scholar at the Hoover Institution, in California, once talked about parties he'd attended that died from a "lethal injection of the American race issue."

I, for one, like my parties like my wine: light and lively. To be cornered and peppered with questions about how "the race" feels about this or that is not my idea of a party, whether the setting is the office or Manhattan's China Club. Anytime we are asked to speak on behalf of the fifteen to twenty million blacks in America, it's off-putting, especially when one's mouth is stuffed with crudités.

Brenda Canty, an outspoken black woman living in Brooklyn with her husband and two sons, says she has learned to respond to the question with a question. "I ask, 'So what do whites feel about so and so?' "

I have taken to using these little opportunities to speak on behalf of the fifteen million blacks in America.

"Well, speaking on behalf of the fifteen or so million of us black folks, we're thoroughly ticked off that whites just can't get over the O. J. thing and go on with their lives," I say with a sly smile. "And if you white folks fail to do so, us fifteen or so million blacks might just take to the streets and riot again!"

My tactic usually solicits a conciliatory grin from the questioner, who by now, I would hope, sees the absurdity of asking one person to speak for millions.

Besides being put on the spot, many blacks will tell you that our idea of a "PARTY" (pronounced PAAR-TAY!) is just not the same as that of whites.

Here's what we see: At white parties people arrive on time prepared to wine, dine, and talk about office politics, child-care problems, summer rentals, and maybe—to liven things up—the local sports teams. They stick around for two hours at most, then they're off to the suburbs to relieve the baby-sitter. Eat beforehand, 'cause all you're going to be served, most likely, is finger food or hors d'oeuvres. When, and if, there is dancing, any black person present is expected to get out on the dance floor and perform for the whites gathered.

Dr. Poussaint once told me that he's not the best of dancers, but whenever he went to a party thrown by whites, the whites in the room assumed he could dance and they didn't understand why he wasn't out on the dance floor.

Blacks, however, don't consider a social gathering a

"PAAR-TAY!" unless there's music and dancing. No respectable black person would arrive at a party on time. Our idea of being on time—and yes, I am speaking on behalf of the fifteen or so million of us—is to arrive at least one hour after the scheduled time, ready to "throw down." (Thus, the term *CP Time,* or "Colored People's Time.") Of course, there's wine and revelry. Of course, we talk about the office, nanny problems, shortage of summer rentals, and sports. But talk is usually minimal. Besides, it's hard to carry on a conversation when the music is at an earsplitting volume and folks are shaking booty in your face. Three to four hours later, we might be ready to wind down, or maybe not. And one expects a *spread*—which means ham, chicken, cheese, crackers, chips and dip, not to mention a variety of desserts. Anything less would be uncivilized.

A case in point.

Paul Delaney, a former senior editor at the *New York Times,* was known as much for his writing and people skills as for the parties he and his wife, Anita, threw at their Upper West Side Manhattan apartment. I must mention here—for it is relevant—that the Delaneys are black.

Paul and Anita's parties were special because they drew a wonderful mix of *Times* staffers, black and white, from every department at the newspaper. Delaney, a man with a wide selection of rhythm and blues, would have this great music playing in the background while we fed our faces with a spread that usually consisted of ham, turkey, and potato salad, among other delicacies.

Unfortunately the music was usually drowned out by conversation. My white colleagues tended to use these gatherings to talk shop and world affairs. The blacks, however, were often guarded, knowing that mixing words and vodka tonics could be lethal, so we kept our opinions to a minimum. We knew that our white colleagues, who routinely arrived early at the Delaneys' fetes, would all be gone by 11 P.M. at the latest; then the real party could start.

It got to a point that we began arriving at Paul's parties later and later to avoid the wine-and-conversation crowd. Well, at one of these parties, one of our white colleagues, a former national correspondent named Jerry Flint, was hanging in there with the rest of the black folks. Finally, all the whites, save Jerry, had departed. We were ready to bump and grind, but Jerry was still haggling. Round midnight, he asked for his coat and was just at the point of kissing Anita on the cheek when he noticed something.

"Wait a minute!" he said. "All the whites are gone. I know what goes on at these parties. You all wait until all the whites leave, then you all start to dancing and partying."

Throwing off his coat à la James Brown, Jerry announced that he was staying and proceeded to the dance floor to party until dawn. He later spread the word, and ultimately there was a group of whites who began to arrive late and stick around for the real party.

Now, I've been to parties thrown by blacks that are as white in nature as any white party and to parties thrown by whites that are as loud, bodacious, and raunchy as those thrown by blacks. But these are usually exceptions to the rule. As such, you will find blacks—be they high school students or working class professionals—who refuse to go to parties thrown by whites.

Private parties thrown by coworkers and colleagues are particularly tricky, especially if said party thrower is "the boss." The black folks in the office know they *should* go lest they be labeled antisocial or be accused of snubbing the boss. But what many whites consider appropriate etiquette at these parties is often a far cry from how we see it.

For example, whites have no compunction about cornering a boss or a supervisor at a party and talking shop or engaging in conversation aimed at making an impression on said boss. They may be having a lively conversation with you, but the minute the boss appears, their eyes look beyond yours, and sooner, rather than later, they make their way to the side of the room where the boss is standing. Blacks—again, many; never all—hate it when whites do that. We are more reluctant to "suck up to the white man."

"It never gets you a damn thing!" said James, a black firefighter from Washington. "They may say something at a party and then act like they were just joking or too drunk to remember. I think black people's pride just won't allow us to do it. Beyond that, it really isn't proper etiquette, and you'd think white folks would know that, since 'they' wrote the book on etiquette."

In Chris Clark's etiquette book, *How to Get Along with Black People: A Handbook for White Folks and Some Black Folks, Too,* coauthored with Sheila Rush, she writes that the social event as an extension of business, politics, or work "is essentially a white phenomenon."

Many blacks have noticed, for example, that whites, especially at work-related social events, try to figure out your credentials—in other words, how you rated to be at this party.

"How do you know Mrs. H.?" I was asked by a white guest during a party in the Fifth Avenue duplex of a socialite hostess, whose name I omit here because of her celebrity. By the tone of his voice, I could tell he'd probably figured out that this little black girl from D.C. and Mrs. Social Register over there didn't go to the same boarding schools, and it's highly unlikely we were presented to society at the same debutante ball, so how'd you get invited to this gig?

My instinct was to say I'd crashed the party. Instead, I said, "We share many of the same interests and friends."

You should have seen the look on his face.

"In black culture the first question at a party might be about where you are from or who your family is," said Dr. Robert Hayles, a black psychologist who counsels businesses on diversity issues. "We do that because you may have people from various socioeconomic backgrounds, and talking about where you work or what you do may alienate those who don't have high-paying, prestigious jobs. Besides that, it's a more graceful way to enter a conversation with a stranger."

And another thing, I've heard blacks complain for years that whites will not patronize black clubs. Now I know what whites are going to say: "Whenever you go into one of their clubs, they [we blacks] give you this evil-eyed look like you're not welcome."

That does happen. But it may come as a surprise that it's usually the exception, not the rule. During the Harlem Renaissance, more whites were seen at the old Cotton Club than blacks. Nowadays whites are reluctant to step foot in Harlem unless they're on a tour bus or with black friends or coworkers.

Don Payne of Houston says whites are often reluctant to be "the only one," yet they expect blacks to feel comfortable in the only-one role.

"Black people will go to a club that's all white, but you will never see a white person at an all-black club," said Mr. Payne. "There's an all-black club here called Thunder Café, and when you go there you can count the number of white folks in there. But go to this all-white club called This Is It, and whites are in there eating and talking it up. When it comes to black establishments, whites go once, if at all, and say, 'Well, we've been there, done that; we don't need to go anymore.'"

Few people I know of like being "the only one." Men don't like being the only one of their gender at a party overflowing with women, although you'd wonder why they wouldn't. Whites say they feel uncomfortable being the only one at predominantly black parties or clubs or schools. Young adults feel uncomfortable being the only one in the presence of older adults. It's human nature, not just a black thang.

And let's fess up here, brothers and sisters. We all know that when a white person is bold and brave enough to break the ice and show up at one of our restaurants or clubs, we resent it.

When more whites began eating at Sylvia's, the popular soul food restaurant in Harlem, black folks didn't like it, even though most of us bit our tongues or our barbecued ribs. But there were cold stares.

We don't have to say anything. We're all thinking the same thing: *Why are they up in here? They got their own places to eat.* (Remember, "Don't want a nigger to have nothin'.")

If whites come to our clubs or discos, we think they're suffering from jungle fever or trying to pick up a few of our dance steps or they're after our men or women.

There may be nothing more at work here than curiosity. Don't we all want to go to the latest hot spot? Don't we all want to eat at places where the food is good and the patrons have some class? What bothers blacks is that too often our establishments don't get the seal of approval until whites show up. Until then, it's just another juke joint.

Mistaken Identities

We've all done it: called an acquaintance, colleague, or classmate by someone else's name. Yet black Americans are more likely to become agitated when whites call them by another's name, because they are convinced the mistake stems from the racial stereotype that "all blacks look alike." It's another form of invisibility.

"We all have gradations in skin color and hair texture, but white people don't seem to make those distinctions," said Cassandra Woods, a thin, dark brown-skinned woman in her forties. A former bank teller, Ms. Woods recalled countless times when regular customers mistook her for another black teller, who was lighter in complexion, heavier in build, and ten years her junior.

"It was bad enough when regular customers did it," she said, "but even some of my longtime colleagues would make the same mistake."

Worse, many blacks will tell you, are whites who tend not to recognize blacks outside their usual surroundings.

"Don't be out of context," said Ronald Prince. "As long as you're at your desk in the office or in the classroom or doing your professional thing, they know you. The minute you're one of the masses, or someplace they think you shouldn't be, you become this faceless blur of blackness."

Whites often mistake me for Whoopi Goldberg. The most embarrassing of these moments occurred on Martha's Vineyard in 1995. I'd been invited by a friend to spend a weekend at her summer home there, and she and her husband took me along to a party at the Vineyard home of a noted white author. I'm wearing a floppy straw hat, baggy dress, and sandals, and I see people staring and smiling and slowly moving toward me. I suddenly get that queasy feeling in my stomach that I'm about to be mistaken for the maid. But I smile back, trying to act like I belong among this mixed group of WASPs, Jewish elite, and black celebrities. The smile must have been some kind of confirmation.

"My son loves you," said one stylish gray-haired white woman.

Before I could say anything, her two friends rushed over.

"Have you bought a house here? I thought you were in Connecticut," one of her friends says, gushing all the while.

"I think you have me mixed up with someone else," I said softly.

The other woman in the group said, "I didn't know you knew the———" she named the host and hostess of the party.

"Well, really, I don't," I said. "I'm Lena Williams. I'm a guest of———" I mention my host and hostess for the weekend.

The smiles slowly reduce to frowns, more of disappointment than embarrassment.

Even some of my black friends say I resemble Ms. Goldberg, whom I've admired from afar. We're about the same height and complexion. We have wide toothy smiles.

But Whoopi's hips are wider than mine, and didn't the Vineyard crowd notice the hair, at least? Do they think Whoopi just cuts and perms her dreadlocks only when she's visiting the Vineyard?

But these were strangers, and their mistake was understandable. It's a far more grievous injury when the mistaken identity is made by whites who, by now, should know better. You know, . . . people you work with, go to school with, live in the same apartment complex with.

"We feel as though we're interchangeable parts," said my brother Ronald Williams, known as Doc to friends and family. "That whites don't take the time to look at us or to get to know us as individuals with our own unique qualities and habits and hobbies. You seldom make the mistake of mistaken identities with people you know well. Think about it. How often have you called your best friend by another friend's name? How often have you called your next-door neighbor by another neighbor's name? These mistakes take place between people who see each other only occasionally, if that."

David Crandall told me in an interview in September 1998 that he no longer took offense when mistaken identities happened only once or twice. What made his temperature rise was how *often* whites made the mistake.

"I'd worked side by side with these two or three white men for twelve, maybe fifteen, years," he said. "They could recognize my voice at a distance because, they said, it was kind of nasal and had a slight southern drawl. But, I swear, three, four, maybe five times a year, they would call me by this other black male coworker's name. Now, although these three white guys hung out together all the time and did sort of resemble one another, I never once called them by any other name than their own."

Mr. Crandall was convinced that blacks are less likely to make such mistakes, because we are used to studying and observing whites from afar. Whites, however, tend to see blacks, he said, as one and the same.

Truth be told, some blacks, myself included, have jokingly played the stereotype to their advantage. A black lawyer at a prominent midtown Manhattan law firm recalled playing hooky from his job to attend a Yankees game with a white colleague.

"He was so worried somebody would spot him in the crowd and couldn't understand why I was acting so nonchalant," he said. "I told him, 'You know what they say: we all look alike.' So what's to worry about?"

I would be less than honest not to say here that blacks are often as guilty of mistaking whites as whites are of blacks. The mistake also rests on the stereotype that "whites all look alike." Same skin color, same hair color, same build.

I remember sitting through a movie, watching Gretchen Mol, the blond actress, and thinking she was Renee Zellweger, the blond actress who played opposite Tom Cruise in the movie *Jerry Maguire*. I've complimented one of my white male colleagues for a story that another colleague, whom the first colleague resembled, wrote. I hate going to restaurants like Hooters or Blondie's, because I usually can't tell my waitress from the other halter-topped blonds. I'm playing to the stereotype, and there's no excuse for it.

But there *is* a difference.

Blacks have been stereotyped for so long, the perception, in most cases, has become a kind of cold, harsh reality. Deep down, many whites truly feel that most blacks look alike.

Blacks have been forced to be observant of white society—their ways and habits—and see life, including our own, through the whites' racial prism.

Dr. Roderick Watts, the black psychologist who lives in Oak Park, Illinois, also sees a difference between black and white stereotyping.

"Blacks are sensitive about generalizations because we believe that the things that are distinctive about us are bad, because whites have caricatured them and thrown them back at us as an insult."

Dr. Watts, who is married to one of my best friends, Isabel Wilkerson, offered this example over coffee in his home:

"You can talk about blacks being loud or blacks being emotionally aggressive. One sounds negative; one positive. Being able to express yourself spontaneously, in an open and free way, sounds positive. There's a certain theatrical, expressive quality about our culture. You can look at it in a negative or more positive way."

Do blacks stereotype whites?

"Yeah, I think blacks definitely stereotype whites. That they can't be cool or hip," he said. "Blacks bark at whites and say things that are rude, stereotypical, and ignorant, but we don't, in a systematic way, deprive them of their livelihood or anything that has an ongoing effect on their life or lifestyles."

First-Name Basis

Among my people, familiarity often breeds contempt. While most Americans enjoy the informality of using first names, many of my black acquaintances turn a deaf ear especially when they are not addressed by their surnames.

"I simply don't respond," said Otis Jenkins, a proud man with an old-fashioned sensibility about right and wrong. In an interview at the midtown Manhattan hair salon where Mr. Jenkins worked as a stylist before he died April 1999, he said the name thing is about "respect denied."

"We're old enough to remember white men referring to black men twice their age by their first names or any name. I saw it done to my father, and his father before him, and I refuse to let them do it to me."

Dr. Poussaint has also felt the sting of familiarity. "When I hear my first name, I immediately react as a black person," he said. "What is this person doing calling me Alvin? They may feel they're being friendly, but you get this emotional reaction."

Roger Wilkins, a former United States Attorney General and currently professor of history at George Mason University in Fairfax, Virginia, says that white people have historically used first names with blacks to establish superiority.

In order to get their due, Mr. Wilkins told David Shipler, author of *A Country of Strangers*, that black people started giving their children names that required white people to address them with respect.

"They'd name their kids Major or General," he said.

Preston King, a black man from Atlanta, Georgia,

took a stand on the name thing and it resulted in thirty-nine years in exile.

In the early 1960s, when Mr. King was a twenty-four-year-old graduate student, the draft board in Atlanta, upon learning that he was black, stopped referring to him in written correspondence as Mr. King. When the board thought he was white, he was referred to as Mr. King. Mr. King told the board he was willing to serve and fight for his country if the board used the proper title in its correspondence to him.

The board refused and Mr. King was convicted by an all-white jury of draft evasion and sentenced to eighteen months in prison. While he was out on bail, he fled to England. This year, some thirty-nine years later, he received a pardon from President Clinton and returned to Atlanta.

My oldest brother, Ralph Williams, who is sixty, says he doesn't take offense when whites call him by his first name, provided he can also refer to whites on a first-name basis.

"But if I'm going to be Ralph and I have to refer to them as Mr. Jones or Mrs. Jones, then I feel I'm being totally disrespected."

As a young civil rights activist during the 1960s, U. W. Clemon was run out of his hometown, Birmingham, Alabama, by Bull Connor, the infamous white Birmingham police chief who used water hoses, tear gas, and police dogs to control peaceful protesters—women and children among them—during the civil rights demonstrations in that city. Mr. Clemon left town that day, but he refused to give up his activism.

Today, Chief Federal District Court Judge U. W.

Clemon of Birmingham doesn't take too kindly to white folks disrespecting him. So when word got back to him a few years ago that an Asian American U.S. assistant attorney in Los Angeles was referring to him as U. W., Judge Clemon wasn't too pleased, to say the least.

"The Justice Department wanted to indict me in connection with the alleged misuse of state and federal funds my sister received for a high school for dropouts she operated in Los Angeles. My only involvement was that from time to time I'd loan her money when she was running low on funds. Anyway, several witnesses who were called to testify before the grand jury told me that this attorney kept referring to me as U. W. during questioning. So I had my attorney write him a letter."

In the letter, Judge Clemon cited the 1964 case of *Hamlin* v. *Alabama*. The case involved a black woman who during the trial refused to answer a white prosecutor's questions because the prosecutor addressed her by her given name, Mary, and not her surname. The judge hearing the case found the woman in contempt and put her in jail. The Alabama Supreme Court refused to overturn the judge's decision and the case reached the Supreme Court.

"The Supreme Court summarily reversed the Alabama courts, and since then the proposition is that you must use a courtesy title during proceedings," continued Judge Clemon. "So we write all this to this prosecutor in Los Angeles. He never wrote back, but I'm told from then on he called me Mr. Clemon."

Whites may see blacks' reactions to such matters as hypersensitivity or our tendency to see race at every turn of a corner. I tend to get far more aggrieved by telemarketers who call my home asking for Lena than my girlfriend Jill does. True, they can't see through the telephone, and the telemarketer could be black, for all I know. But I think they know I'm black because...well, maybe I sound black. (We'll address that little stereotype later.) Besides, most blacks know not to call a person's home and refer to the person on the other end of the receiver by first name, even if the boss says that's how it's done in the telemarketing world. There is a right way of doing things and a white way of doing things. The latter is the white way.

So whenever they come a-calling all friendly like, asking for Lena, I ask them if I know them.

"Well, dearie, only my friends call me Lena. To everyone else, I'm Ms. Williams. Now, seeing as how you've already disrespected me by calling me out of my name, you really don't expect me to be interested in anything you have to sell or tell me now, do you?"

I know I may be overreacting, but boy does it feel good when I hang up the phone on the telemarketer who, by now, is mumbling effusive apologies.

While I'm on the subject of names, there's another unconscious habit many whites have of taking liberties with people's names, by either abbreviating the name or using nicknames.

Whites who do this may be guilty of nothing more than trying to be friendly. But where I come from, you just don't go around calling every Robert, Ronald, Susan, or Elizabeth...Bob, Ron, Sue, or Liz. Not unless it's OK with the person.

Most blacks who know Gerald Boyd, a fellow black journalist and *Times* colleague, know not to call him Gerry. Gerald doesn't like being called Gerry. Yet when he was assigned by the *Times* to cover the Ronald Reagan White House, President Reagan, being the kind of folksy leader he was, routinely called Gerald, Gerry. Out of respect to the man and deference to the office of the presidency, Gerald never told Mr. Reagan that he really didn't like being called Gerry. But boy, did the rest of us kid him about it. Now, as deputy managing editor of the *Times*, I try my best to get under Gerald's skin by calling him Gerry.

"I think it's a white American thing," said John Tate, another old family friend. "They may not mean anything by it, but blacks see it as taking liberties that weren't necessarily granted. I wouldn't really like it if someone called me Jack, which is a nickname for John, because my parents didn't name me Jack."

Once in grade school, I answered a question incorrectly simply because I didn't know that John Fitzgerald Kennedy and Jack Kennedy were one and the same. That was the last time I made that mistake. From then on I learned important nicknames because I figured it was important to whites, so it'd better be important to me. You weren't going to get a Theodore Roosevelt by me—Teddy. Dwight D. Eisenhower—Ike. Richard Nixon—Dick. Abraham Lincoln—Abe.

I later learned that these little name games weren't being played only by whites.

"Aunt Lena, you have to give us a heads-up on your friends," my nephew Frank Jr. told me one day.

"What do you mean?" I asked.

"You just tell us that one of your friends or colleagues is coming to D.C. and may look us up, but you don't give us any idea whether to expect someone black or someone white, so we're here trying to figure out the race by the name."

He told me that they—mainly he and his cousins—figured Jill would be white. (Right.) Judy was white. (Right.) Diane? Black. (Right.) Tyrone was black. (OK.) Leroy would definitely be black. (Right.) Ronald was black. (Right.) But Ron would be white. (Well, all right!)

This little name game my nieces and nephews were playing on me got me to thinking about a little thing that hadn't come to mind. Experts who study names say there are racial, ethnic, regional, class, and generational patterns to names. Not only is Beulah most likely to be black, she's probably going to be a woman now well into her late sixties or early seventies. Similarly Eve is likely to be an older white woman. Ruby is more likely to be a black woman born after World War II, while Amber is most likely to be a white woman born in the late 1960s or 1970s. The Tamikas, Shantices, and Maliks of the world are likely to be blacks born in the 1980s, when a younger generation of black parents dared to be different by giving their children names that they assumed would defy generalizations and set them apart from the Ebonys and Africas of the previous generation.

It didn't quite work out that way. Whenever I hear of a Tyese or Kimara, I automatically think black child.

"You don't find a lot of blacks with cutesy names like Amber," said John Ekizian, my friend and agent, who is white. "And I don't think I've ever met a white Leroy. But I have a friend named Thomas Jackson who is white, and you can't believe how many people think he's black."

Even nicknames have racial patterns.

Buffy, Bif, Whit, Winny, Gidget, are just a few nicknames popular among whites. June Bug, Jitter Bug, Junior, Little Bits, Slim, T, Slick, are black nicknames that seemed to have survived generational trends.

So what's in a name? you may ask. For many blacks, a painful history.

Our black ancestors were given their surnames by their English-speaking slave masters, who needed a way to identify their property. During the black-power movement of the 1960s and 1970s, blacks were encouraged to drop their slave surnames and assume a more Afrocentric or Islamic identity. Some dropped their surnames altogether. The Nation of Islam chose X. Many blacks kept their original surnames but changed their given names to reflect their African ancestry, thus Jamal, Kenyatta, Chuma, Nairobi, et cetera.

But history wasn't what my nephew was talking about when he mentioned getting a heads-up. Frank explained that the reception he gives my friends doesn't differ dramatically because of their race, but he, like so many other blacks, often acts or behaves differently around blacks than he does around whites. No matter how close we may be to them, we do everything from using the King's English to not speaking ill of other members of the race.

"How often do you hear blacks say, 'We talk one way at work and another way when we're hanging out with friends in the old neighborhood'?" said Frank Jr., who teaches music at an elementary school in Maryland. "I think we—blacks, that is—feel freer to be ourselves around our own people than we do with whites, who may misinterpret our actions."

By knowing what race belongs to a name, whites also can avoid "the look," like the one Aaron Walton, a black executive sometimes gets when he arrives for a meeting and whites see a black man standing in front of them.

"The name sounds Jewish," said Mr. Walton by way of explanation.

Experts are divided over whether a name can predict or affect future achievement. Blacks are largely of one mind on the subject. Many blacks worry that as names become readily identifiable by race, black children will be stigmatized or subjected to other people's prejudices. When a Zaire or Ashakulu applies for a job or admission to an exclusive school or club, it could make it easier for those who are prejudiced to consciously or unconsciously discriminate.

6. Little Things in the Mass Media

The Negro Problem

A FEW years ago the Pulitzer Prize–winning editorial cartoonist for the *Chicago Tribune* created a cartoon that depicted Goldilocks peering up into the eyes of Papa Bear and innocently declaring, "It was a black guy!"

The cartoon appeared in 1994 shortly after Susan Smith, a white mother from Union, South Carolina, claimed that a black man had abducted her children in a carjacking.

When the car was later fished from a river with the two dead boys inside, still strapped into their car seats, the nation gasped collectively in shock and horror, and black men found themselves under siege by the media and by law enforcement officials. Before a lynch mob could gather, Mrs. Smith confessed. Spurned by an ex-lover, the petite sandy-haired mother, who married her high school sweetheart, had taken the lives of her toddler-aged sons and blamed the crime on a black man.

While whites were horrified that a mother would kill her own children, blacks were equally disturbed that white America was so quick to believe that "a black guy did it!"

"The black bogeyman," said Darren Grinage, my twenty-six-year-old nephew. "They blame us for everything—crime, drugs, carjackings. I'm surprised they didn't put El Niño on us. If you believe what you've seen or read in the mass media about blacks, we're a violent people, you know."

Whites say—and in some respects probably believe—that blacks are responsible for the majority of crime in America. Although a disproportionate percentage of crime is committed by blacks compared to their numbers in the general population, most crimes committed by blacks are against other blacks, not whites.

What disturbs many blacks, especially black men, is that the "white lie" Mrs. Smith told that day fits a long history of the scapegoating of black men, from the Scottsboro Boys, the nine black teenagers who in the 1930s were falsely accused of raping two white women to William Bennett, the black man who was arrested and charged with shooting a white man and killing his pregnant wife during a carjacking in Boston in 1989.

The Stuart case, as it became known, struck fear into the hearts and souls of black folks. Boston, a city not known for its racial tolerance, was close to civil war the day Charles Stuart, a white businessman, told police that a black man had shot him and killed his wife as they drove from a childbirth class. Dozens of black men were searched and questioned by the police. And Mr. Bennett was arrested and imprisoned before Mr. Stuart's story began to unravel beneath a tale of adultery and financial fraud. Before he could be tried, Mr. Stuart jumped to his death into the Boston Harbor.

I don't want to sound cold or callous here, but he could have saved a whole lot of black men considerable grief had he jumped sooner.

How many white males who bore some resemblance to Theodore Kaczynski, the Unabomber, were subjected to humiliating searches on busy streets or in the privacy of their homes? Mr. Kaczynski was able to evade police authorities for several years, killing and maiming scores of innocent citizens during that time. But one black guy on the loose and it's time to hide the women and children.

A United States Court of Appeals for the Second Circuit, which includes upstate New York, said that police officers did not violate the Constitution when they stopped every black man in town to find an alleged robber. That's exactly what police officers in Oneonta, New York, did in 1992 after a seventy-seven-year-old white woman said she had been robbed in her home by a young black man.

To find the suspect, the cops stopped every black man they encountered in the small upstate New York town. Of the fourteen thousand residents, fewer than five hundred are black. No reason to narrow down the suspect list when you're only looking for five hundred usual suspects who fit the description of "black."

So the police went out searching for their man. They stopped black men on street corners. They pulled them over in their cars. They pulled them off bicycles. They went to the State University of New York at Oneonta and got a list of all the black students in the school and went looking for them. One cop even stopped a black woman.... Don't ask!

The cops never found the alleged "perp."

A group of blacks in Oneonta filed suit, claiming the police had violated the equal protection clause of the Fourteenth Amendment and the Fourth Amendment's prohibition of unreasonable search and seizure.

The federal appeals court didn't see it that way. The court ruled that the blanket dragnet was not racially discriminatory, because the cops were acting on a description that included more than just the color of the alleged assailant—the woman had told police she never saw the man's face but could tell from his arm and hand that he was black. She had also said she thought the assailant was young, because he moved quickly.

Gives new meaning to *blind justice.*

Stereotypes may begin at home, but they are often spread through the mass media, white-owned institutions that help shape our opinions, views, and understanding of each other.

The mass media representations of welfare mothers as single black women and of black males as the embodiment of drugs, disease, and crime are so dominant in American culture that most white folks accept them as fact—truths that are self-evident. Round up the usual suspects: black males, medium build, in their late twenties or early thirties. We're only talking a few million folks, people!

In a debate at The Apollo Theater, in Harlem, between Vice President Al Gore and Senator Bill Bradley of

New Jersey, the two Democratic candidates for president were asked, in light of cases like that of Amadou Diallo, what concrete steps they would take to deal with police brutality and racial profiling, without increasing crime.

Senator Bradley said he felt the shooting of Mr. Diallo was "an outrage" and "a tragedy."

Then Bradley, whose appeal in black circles dates back to his basketball-playing days with the New York Knicks, spoke volumes: "But I also think it reflects racial profiling in the sense of racial profiling that seeps into the mind of someone so that he sees a wallet in the hand of a white man as a wallet, but a wallet in the hand of a black man as a gun. And we, we have to change that."

If elected president, Bradley pledged to issue an executive order that would eliminate racial profiling at the federal level and to get a law passed to get information gathered at local levels so that "we can see how the police departments are acting."

"I would make sure that the Justice Department was involved and would say quite clearly that white Americans can no longer deny the plight of black Americans."

Count it and add one! (Sports jargon for counting the basket and getting a free throw, to boot.)

Not to be outdone, Gore said that America has to recognize that racial profiling "is a problem not only in law enforcement but also in insurance, in banking, inside schoolrooms, and inside people's hearts."

As a journalist of some twenty-eight years and a member of the National Association of Black Journalists, I've tried to present a more balanced picture of black Americans and the way we live. I try not to distort the truth. When blacks do or say something wrong, I'm the first to note that. I have never subscribed to the mantra of "my people, right or wrong." But there have been times when I have refused to write stories that seem bent on fueling black stereotypes. I mean, how often can we psychoanalyze Mike Tyson? Do we really need another series on life in the inner city? And can we cease with the criticism of the "overpaid, sexually irresponsible, hostile black athlete"?

The six o'clock television news often begins and ends with images of black youths: The top of the news hour gives us the daily dose of black youths in hooded sweat-shirts, heads down, being marched, handcuffed, into police precincts; the bottom of the news hour features young blacks slam-dunking their way out of poverty.

To illustrate that point, Carol McRae asked this rhetorical question of those gathered for the focus group in Washington: "How many of us have said that if the news fails to show a picture of the suspect in the first five seconds of the report, then he must be white?" She looked around the room to nods of affirmation. "If the suspect is black, you're going to see a photograph of him splashed across the newspaper or the six o'clock news."

Too often the media prefers the stereotype to the authentic black experience. After a while whites don't know the difference. The original Bill Cosby television show was criticized for being "unrealistic" and "not black enough." Too much father-knows-best and less father-could-care-less, I guess. What's unrealistic about a black family where the father is a physician and the mother a lawyer? I know a few black families like that. What's "not black enough" about blacks who appreciate the life, culture, and heritage of their people? The Cosby show was criticized because it refused to play to the stereotypes, while shows like *In Livin' Color,* and *Martin* are celebrated by media critics for doing just that. Black men in drag, pistol-packing mamas, hoods in the hood, and female dancers who can bump and grind with the best of 'em—now, that's an authentic picture of black life in America.

Think of it this way. If you're E.T., the extraterrestrial, and you've just landed on Earth from outerspace, based on what you see on television or in the movies and what's in the newspapers or newsmagazines, what would you think about black people? That we're criminals, athletes, or single women with children? Oh yeah, and then there are these three exceptional black folks called Colin Powell, Michael Jordan, and Oprah Winfrey.

Even when the white media tries to present positive black images, there can be problems.

In April 1998 *Boston Magazine* was shocked that black Americans took offense at the headline on the cover: HEAD NEGRO IN CHARGE. The article was about the nationally known scholar Henry Louis Gates Jr., chairman of black studies at Harvard University.

The NAACP and the Urban League complained to the magazine that the term is racist and harks back to the days of slavery. The magazine's editor, Craig Unger, said that he was "genuinely upset if people saw this as racist."

"That was not our intention," he said.

Aaah, the old question of intent.

"Your honor, I didn't mean to kill 'em. It just happened."

The defense doesn't work in court, and it usually doesn't work with us black folks.

The group of white men from Queens, New York, who were videotaped in blackface and Afro wigs, riding atop a float during a parade in the summer of 1998, sporting signs that declared "Black to the Future," were "just joking around."

I guess they were still joking around when one of the men decided to dangle from the back of the float an effigy of James Byrd Jr., the black Texas man who was dragged to his death in 1998 by three white men.

Speaking in their own defense, the men said they'd ridiculed Arabs and Chinese Americans in previous years.

I know what many whites will say: Blacks can dish it out, but they can't take it. Point taken.

There is a fine line between satire and cruelty, and shows like *In Livin' Color* tiptoed across that line as if it were a tightrope.

No group was immune. Not gays, not the elderly, not the disabled. Or Hispanics or Asians. And for sure, not whites.

"I used to watch that show, and for the life of me, I never understood how they could get away with some of the things they did," said Ryan, a young white man who happened to be staying at the Hyatt in Bethesda, Maryland, when I was holding the focus group there, and he decided to see what it was about. "What surprised me was that the show depicted blacks in very stereotypical ways, but blacks seemed not to mind at all. From what I understand, the show was very popular among blacks. But I know a white comedian couldn't get away with doing that."

Was he bothered by how the show treated whites?

"Kind of," Ryan acknowledged. "Black comedians like Richard Pryor and Chris Rock always made fun of whites. Some of what they joke about is true, some isn't. But I never considered anything they said to be cruel."

Not that anyone would care if he had. It's open season on white Americans, especially white males.

Ishmael Reed, the black novelist and social critic, has described white men as "the biggest minority group in America." According to Reed, everyone in America can dump on white men and no one empathizes.

Whites don't complain, because they know their complaints will fall on unsympathetic ears. But let a white man, be he ordinary citizen or comedian, say something in jest and he's going to be held accountable.

Fuzzy Zoeller's finger-snapping comment about fried chicken and collard greens being served at Augusta Country Club after Tiger Woods won the Masters Tournament in 1997 may have played to a stereotype, but it was rather funny, folks. But Zoeller found that the only things *he* was eating after that were his words.

Race has made us a nervous country, afraid to laugh, and I don't know why, because some of this stuff is so hilarious.

The Games White Folks Play

One of the most popular shows on television in 1999 was the game show *Who Wants to Be a Millionaire.* Contestants on the show, which was hosted by Regis Philbin, had to answer a series of questions within a given time frame. For each correct answer, the contestant received cash, accumulating up to one million dollars. The show was praised as innovative, not only because it offered a new take on the game show genre but also because it featured a diverse group of contestants, including openly gay couples.

Although there were black contestants, the show's producers were surprised that so few blacks auditioned for the show. Before being selected to the show, contestants had to pass a series of behind-the-scenes tests of their knowledge. Despite efforts to broaden the pool of contestants, black Americans seemed to shy away from what appeared to be a golden opportunity to become an instant millionaire.

You may wonder why.

The myth that blacks are intellectually inferior to whites has persisted despite our best efforts to debunk it. Whites will seize upon any opportunity to prove themselves superior. Though the show may have been fun and games to white America, blacks would have had to take it dead serious. Besides that, most of the questions dealt with trivia—not exactly black Americans' strong suit.

We've spent so much of our lives trying to learn the white man's English, the white man's history, the white man's ways. Who has had time to deal with trivia? Now, if knowing the name of the sled in Orson Welles's film *Citizen Kane* was going to land us that government job, it might be a different matter.

In the 1960s—that not-too-distant past—blacks in the South had to pass literacy tests in order to vote. We all remember the stories about blacks who could recite the preamble to the Constitution, could name every United States president, knew by heart the signers of the Declaration of Independence and the capitals of every state in the union, only to be denied the right to vote because they couldn't guess the number of jelly beans in a glass jar.

In the 1960 novel *Black Like Me,* John Howard Griffin, a white man who used medication to darken his skin so that he could live as a black man, wrote about an incident that took place in Alabama when an older black man tried to register to vote in a local election.

The man correctly answered every question he was asked. Frustrated, the white tester handed the man a Chinese newspaper and asked if he could read it. The man said he could. The white tester, turning redder by the second, asked the black man what it said.

"That this is one nigger that ain't going to vote today," the man answered.

We must know everything there is to know about matters of importance to whites. Do you honestly think we have time for trivia? Besides, most of us—add me to the list—think those game shows are rigged. They're not about to let one of us show them up on national television and leave the viewing audience with the impression that a black person could actually be smarter than a white person. You won't be a millionaire at their expense.

The Way We Were

Nostalgia can be a good thing. We all recall bygone days when life seemed so much simpler.

At some point in our lives we've all longed to return to our glory days, when entire communities did indeed help raise our children, when people could leave their front doors unlocked, when teachers were treated with the utmost respect, when a proper education was viewed as an avenue toward achievement.

But some blacks see these moments of nostalgia as longing for a return to when whites reigned supreme and blacks knew and stayed in "their place."

When whites engage in these trips down memory lane, many blacks become agitated. We respond instinctively to an inherent assumption on the part of whites that the past is preferable to the present, and to the future: when minorities will be the majority in America.

"Whites don't want to think about that possibility," said H. T. Starr McCauley, who is black. "They can't fathom a time in this country when they will be in the minority. They look at America today and see a country in chaos. Nowadays, black athletes are dominating various sports. There are blacks making as much money, and more money, than whites. There are blacks in powerful positions. Blacks married to whites. You didn't have that in the good ole days, because whites wouldn't allow it. So, all this longing for the past is nothing more than a return to the status quo: when white was right!"

This fundamental belief that America was better-off years ago sometimes irritates blacks to the point of irrationality. Or is it irrational? Consider this:

I've had black women who own furs say that they do not support the anti-fur movement because as long as white women were wearing furs in the 1950s, 1960s, and 1970s, nobody seemed to have a problem with it. But the moment black women were able to afford furs, all of a sudden furs have to go.

"It's a movement started by whites and whose members are predominantly white," said my sister Gloria, a black woman who has a sassy mink to match her attitude. "They want us to burn our coats because trappers shouldn't be out there killing little minks in the northern woods. Well, white cops are out there killing little black boys right here in the inner cities and I don't see white folks throwing red paint on the precinct's steps."

This feeling of old versus new can also be seen in the sports arena. *When It Was a Game* is a popular video about baseball in the days of Ruth, DiMaggio, Gehrig, Mantle, and Maris. The days when baseball was the "true" American pastime. Nowadays America's pastime seems more like Latin America's, with players like Benitez, Gonzalez, Hernandez, and Sosa.

Walk into any sports bar and you'll hear older white males waxing nostalgic about Ruth and Gehrig, as if baseball died when they did. Robinson, Mays, Aaron. These black athletes also played the game with as much enthusiasm, flair, respect, and love as any white player. What I, as a black American, hear in such conversations is racial doublespeak. An underlying desire to return to the days when "even the ball was white," when our heroes and role models were white, when whites were hip and actually could dance!

Say what you like about that issue, but the kids on *American Bandstand* could stroll and bop and not miss a beat.

We can never go back and, truthfully speaking, blacks aren't the only ones celebrating that fact. The past wasn't exactly a blast for women, either, white or black. We won't be returning to the kitchen, thank you. We might as well move on and let the past be a memory.

You Know What They Say about Them

"You didn't like *Waiting to Exhale*? a white female colleague asked, her voice tinged with an air of incredulity.

"Frankly, no," I answered, not hiding my annoyance.

"Why not?" she wanted to know.

"Because it portrayed black women, no matter how high we rise in the professional world, as hateful, foul-mouthed sluts who can only be made whole by having a man. And"—I continued, never pausing to take a breath—"not every black man out there is a low-down, skirt-chasing scoundrel. But you wouldn't have known that by reading the book or watching the movie."

My colleague shrugged and returned to her desk.

In retrospect I'm not sure why I responded with such passion to her question. But it didn't take long to figure out. I, like many other blacks, resent being expected to like something simply because it's about black Americans or because white America has sanctioned it as "fine art."

Terry McMillan's 1992 best-selling book about the trials and tribulations of five single black American women was a popular work of fiction that apparently struck a chord across racial and, in some regard, gender lines. The movie *Waiting to Exhale,* starring Whitney Houston and Angela Bassett was released in December 1995. Black women lined up to see the movie, making it one of the top-grossing films of all time. From what whites could see, "all the blacks they knew" loved it. So what was wrong with me?

What was wrong was that movies like *The First Wives Club* could convey white women's indignation over their ill-treatment by men and do so without the women being portrayed as bitches or whores.

In retrospect, I wasn't simply responding to the movie. That was a little thing. There was a larger thing simmering in my soul.

As a single black woman, I've had it with white folks' pity about how difficult it must be for attractive, intelligent black women over thirty-five to find suitable mates. "We just want all of you to exhale and be happy!" Right.

The wives in *First Wives* may have been dumped for younger women, but us poor black women can't even find the first man to marry. Now, I've had black friends, male and female, express empathy over the disproportionate percentage of eligible black women to eligible black men. The older we get, the greater the disparity. Factor in age, sexual orientation, the number of black men in jail or dependent on drugs and those who simply aren't suitable mates for either social or economic reasons, and our prospects become even dimmer. Blacks know this. We've bemoaned the reality and debated ways to fix it. Yet never once have I felt that blacks who share my pain do so in a patronizing or condescending way. Whites, however, even if unintentionally, tend to view these facts of life as some quirk of nature—a malaise that afflicts black women only, leaving us embittered and frustrated.

I've had whites ask, perhaps out of sincere concern, what's wrong with the black men out there, referring to the footloose, fancy-free ways of our men. It hurts. Although studies have shown that single white women over thirty-five also face difficulty finding candidates for marriage, they are not confronted with the same obstacles placed before black women. And whenever I see a black man with a white woman, I see one less eligible black man for a sister. But I'll deal with that issue in another chapter.

Isn't He Articulate

If I counted the times a white person—whether television commentator, journalist, or employer—used *articulate* to refer to blacks who use proper diction or English in expressing themselves, I might be left with little else to do.

A white colleague of mine, Clyde Haberman, told me that he once fought vociferously with an editor who wanted to describe members of the City University of New York's debate team as articulate.

"They'd just defeated Harvard's debate team," explained Mr. Haberman, "and I felt that said enough about how the team handled themselves."

"I get the articulate thing all the time," said Mr. Frazier of Washington, who has a masters degree and, thus, should be articulate. "The assumption is blacks can't speak, can't write, and can't think. At my job, we have a very high writing standard and upper management will correct written mistakes. One day, by mistake, I received a stack of correspondence that contained some of the white manager's work, and there were more red marks and corrections than in mine."

Upon hearing that blacks take exception to the frequent use of that term, my friend Jill said she began to think twice about using it in some of her news articles.

"It just never occurred to me, and I use that a lot," she said.

"That's not the problem," I explained. "It's fine to say a sixth-grade student was articulate. But one would assume that it goes without saying that Colin Powell would be articulate; so why say it?"

The greatest offenders regarding the articulate thing are probably sportscasters and sportswriters. And yes, I am a member of that club; but I know better.

"Oh, isn't Michael Jordan articulate?"

Of course. He's being paid millions of dollars in endorsements and promotional fees because he's articulate. AND, he's good-looking!

Ever notice that sportscasters or writers don't say that about Larry Bird? Now, I'm not going to go off on a Larry Bird trip. I've always appreciated Bird as a player, and from what I've heard over the years, he never allowed himself to be baited by the race debate. But with all due respect to Bird's outside shot, the man isn't exactly articulate. Yet, you've probably never once heard anyone say that, at least not out loud.

Having started my career as a sportswriter, I know firsthand the daily trespasses among the media and black athletes.

Racial slights in the sports world are plentiful. And our memories of slurs, derogatory comments, stereotypical portrayals, are long. Blacks couldn't forgive Jimmy "the Greek" Snyder for his comment to a television reporter that the superior skills of black running backs and sprinters versus that of white runners had to do with breeding.

Mr. Snyder had a theory as to how this physiological difference came to be. During slavery the "big black buck was mated with the black woman." This pairing resulted in offspring with high behinds and muscular legs. Jimmy "the Geneticist" Snyder lost his job as a commentator for CBS *Sports* for that little remark.

Many whites accused blacks of overreacting to a blatantly ignorant comment made by a man who'd had too much to drink at a popular Washington, D.C., bar. Hey, we didn't tell CBS to fire the guy. Indeed, many of us had a good laugh about it! Jimmy "the Greek" only said publicly what we knew many whites were saying privately. Mr. Snyder apology, made immediately after his comments were aired across America, would have sufficed. But CBS, in what many blacks viewed as a hollow attempt to demonstrate their sensitivity toward matters of race, removed him from the air.

Al Campanis, the Los Angeles Dodgers' general manager faced a similar fate when he, also in a televised interview, attributed the dearth of blacks in front-office positions in baseball to a lack of skill. "Blacks can't swim because they lack buoyancy," Mr. Campanis noted, by way of example. He, too, was fired.

Whites saw these dismissals as a nation bent on political correctness. But there were many prominent blacks—politicians and clergy among them—who felt anything less would have been a miscarriage of justice.

Most of the blacks I interviewed were not at a loss for examples when it came to slights uttered about black athletes.

"If a black excels at a given sport, he's called 'a good athlete,'" said Al Harvin.

Now retired from the *Times,* Mr. Harvin, a Harlem resident, continues to write on the subject. "Whites are referred to as 'smart players.' Commentators applaud white athletes as 'hard-workers' or players who 'hustle.' Black players? Well they have 'natural ability,' as if natural ability flows freely with little or no practice or honing of skill and talent."

Sports have long been my salvation; my escape from reality. I can sit in front of a television set and watch game after game: football, basketball, baseball, hockey, tennis. Then, someone comes along and injects race, and it's a whole different ball game.

During the 1998 baseball season, the country was riveted by the home-run race. Mark McGwire, the first baseman for the St. Louis Cardinals was threatening to topple the sixty-one home-run season mark set by Roger Maris of the New York Yankees in 1961. Sammy Sosa, the right fielder for the Chicago Cubs, was also on a pace to break Maris's record. McGwire is a tall redheaded Irish Catholic from an upper-middle-class California family. Sosa is a compact dark-skinned man from the Dominican Republic, who shined shoes to help his impoverished widowed mother earn a living for his eight brothers and sisters.

Across America everyone talked about the race and which of these two talented athletes would be the first to sixty-one. I was pulling for McGwire, a muscular, home-run hitting machine who—during an illustrious career beginning with the Oakland Athletics—hit fifty or more home runs in two seasons and forty-nine in his rookie year. From what I'd read about him over the years, he seemed like one of sports' good guys: modest, unassuming, an all-around team player. McGwire had come three home runs shy of the record sixty-one mark in 1997. Here was an athlete who'd handled with aplomb the media pressure from the start of spring training in March.

Sosa had come into his own during the 1998 season, when he hit a record twenty home runs in a single month (May). The previous year, Sosa had held the record for strikeouts at the plate.

The race thing surfaced around the time these two fine players both had fifty home runs, by the end of August, a mark that put them on a pace to hit well beyond sixty-one.

Some members of the media suggested America was

pulling for McGwire because they wanted "an American" to break the record. Others said they wanted a white player to break the record. McGwire and Sosa never responded to these assertions.

But there were hints of racial preference. A white female friend, who is a die-hard New York Mets fan, told me that she was at a Mets game that September, when the Cardinals and Cubs were playing in St. Louis. At the time, McGwire had sixty homers, to Sosa's fifty-eight. When McGwire hit home run number sixty-one to tie Maris, my friend said a group of young white men around her cheered and gave each other high fives. They also cheered each time Sosa struck out.

"Maybe they were jeering Sosa because the Cubs and the Mets are vying for a wild-card play-off berth," I suggested. I knew better, but I didn't want the moment ruined by racism's ugly head.

"Naah," she said emphatically. "They showed McGwire hitting his sixty-first on the screen so everyone in the stadium could see it. After that, they just showed the score. But one of these guys was still listening to the game on the radio, and whenever Sosa struck out, he announced out loud: 'Hey, he struck out again.' And they'd all giggle and cheer."

For that one moment, my white friend understood the nuances of racial prejudice that I, and other blacks, have learned instinctively to detect.

Miss Scarlett, I Don't Know Nothin' 'bout Birthin' Babies

The feeling that whites fail to recognize the importance history has played in the lives of black Americans is experienced in ways that make even ordinary occurrences seem like major transgressions.

In the summer of 1998 the American Film Institute asked Americans in the film industry to choose the top one hundred films of all time, using as criteria the films' awards, gross, content, among other things. Included on the list were *Gone with the Wind,* released in 1939, and *Birth of a Nation,* released in 1913.

In terms of cinematic scope and longevity, one would be hard-pressed to argue the films' selections.

But many blacks, young and old alike, fail to see the redeeming value in either film. How could they? In both movies blacks are portrayed as subservient, amoral human beings. *Birth of a Nation* fueled the stereotype of black men as conniving oversexed beasts who lusted after white women. While white America holds these two films in reverence, we would consider our culture enriched if the films had never been made.

Another considerable source of irritation for many blacks is the perception that whites fail to support our historical institutions.

"They built a Holocaust museum in Washington and called it the United States Holocaust Museum," my niece pointed out one day. "What has America done to institutionalize our history in this country? You won't find it, because whites don't want to own up to the atrocities they inflicted upon us. And how many whites have visited the Birmingham Civil Rights Institute or the African-American Museum in Detroit?"

While researching this book I visited the Civil Rights Institute and African-American Museum to see for myself if her suspicions bore fruit. On a sunny Saturday afternoon in early June, I went to the Birmingham Civil Rights Institute with one of my best friends, Ronald Prince, a native of Birmingham. We chose a weekend day because it offers a different barometer by which to gauge public support. The number of visitors to museums during the week is often inflated by students visiting as part of a classroom project. However, people who take the time from their weekend to go to such places are more likely to be doing so because they want to go, not because they have to.

There were no more than a handful of tourists at the institute when we arrived. But in the crowd were two white women. I didn't ask why they'd come, partly because I did not want to put them on the spot. That they were there was testimony enough for me. By the time we began the hour-long tour, we had been joined by nearly forty others—all black.

In a way I was disheartened. I went back to the institute during the week, searching for evidence to prove me wrong. It was not there. What I did find, however, was a group of twenty predominantly white college students who were studying the civil rights movement by traveling through the South and speaking with some of the leaders of that movement.

The students—from five colleges, including Bennett and Marietta—had come to the institute that day to hear four black Birmingham residents who had been a part of the movement talk about the struggle for freedom and justice.

"Have you ever received an apology from those who did this?" one white student asked Reverend Franks Dukes, who marched with Dr. King and was once jailed for his civil rights activities.

Another white student noted that while the laws may have changed, the beliefs of whites in the South may not have changed with them. "Are biases still there?" he asked.

"There is still a modicum of racism in the South," said Arnetta Streeter Gary, who, as a sixteen-year-old eleventh-grade student, had marched to the Birmingham courthouse one day with other protesters to register to vote and was jailed for three days for her activities. "It's taken on different attire now. People are more educated now. If they are prejudiced, they are educated enough to know it's not a positive thing."

Ms. Sherwood, a white postal worker who was raised in western New York and now lives in Rockville, Maryland, said most whites grew up believing what they read in history textbooks, which, for the most part, omitted black history.

"They don't know," she said. "I had an uncle who taught history at a university. When I was in high school, I read a lot of history books. Back then the history books left you with the impression that slavery wasn't that bad. Life on the plantation was OK. The slaves were fed and clothed.

"But when I said that to my uncle," Ms. Sherwood went on, "that slavery wasn't that bad, he gave me a copy of the writings of Frederick Douglass to read. That's when I began to learn about black history and black culture and black pioneers like Charles Drew and Madam C. J. Walker. Now I know about the 'big lie'—the one about blacks not having contributed to America's history."

What's Up with That?

Nowadays Negroid features—full mouths, high behinds, dark skins—are in vogue, and some whites, women in particular, are resorting to silicon injections to get fuller lips and firmer butts. Now that white cover girls have a more voluptuous look, "it's sexy." Now it's beautiful.

But whites will never say that these cosmetic changes are efforts to "look black." Nooo! Words like *exotic* and *ethnic* have been used to describe the trend, but nothing that connects it to black culture.

And you know another little thing I've never been able to figure out for the life of me? How whites can publicly denigrate our black skin, yet spend hours in the sun trying to get darker. Never mind the risk of skin cancer. Why would you want to look like a people you consider inferior in every respect? Blacks weren't the ones calling whites paleface, at least not publicly, anyway. Whites referred to each other that way. I can recall a popular 1960s television commercial for Coppertone suntan lotion: "Don't be a paleface. Tan—don't burn. Get a Coppertone tan."

"What's so irritating is that they think their suntanned darker skin automatically makes them beautiful," said Barbara George, a Manhattan hairstylist. "Actually, it makes their skin look like rubber, but you didn't hear that from me. But we're supposed to think that our naturally, God-given dark skin is unattractive? Then, why are you trying to get darker skin, huh?

"And you know something else whites will do that really annoys me?" she went on. "They act like we can't get suntanned or sunburned. I remember once going to the Caribbean and coming back home a shade darker and having one of my white neighbors say, 'You got a suntan?' When I told him I did, he said that he didn't know black people could get suntanned. And I said, 'Well, can white people get frostbitten?' He didn't know where I was going with it, but I knew. I told him that it's widely believed that whites love cold weather, that they can stand more cold than blacks can, and that blacks can stand more sunshine. Well then, it would stand to reason that if blacks have so much melanin in their skin that they, for some reason, can't get suntanned or sunburned, then it would stand to reason that whites have so little melanin that they don't get frostbitten."

Whoomp, there it is!

My niece Antoinette often gets asked about her freckles.

"I've had whites ask me whether I was out in the sun, because I have freckles," said Toni of the brown freckles splattered across her nose and cheeks. "I say, 'No.' They act as if blacks aren't supposed to have freckles. They also are shocked that our hair gets lighter if we're in the sun for long periods of time."

And what's up with black slang? . . . No, not Ebonics. That's black dialect—slave dialect with an attitude. "Like we b' gonna do somein."

Naah. I'm talking about words or phrases often used by a younger generation to alienate the older generation. Now, I know there's no copyright on words or popular phrases. Slang is as American as apple pie. Generations can be defined by their use of slang: *cool, neat, hip*—50s slang; *dig it* and *sock it to me*—60s and 70s slang; *go for it, make my day*—80s slang; *whatever, don't go there*—90s slang.

So how come, in the mouths of black youths, slang becomes a lethal weapon?

"Yo, homeboy, what's up?"

"Man, that's the bomb."

"Chillin' and illin'."

Whites can hear these innocuous comments uttered by black kids on the street or public transportation and try to make a case for Ebonics in school curricula.

Then, when the so-called urban vernacular becomes mainstream and everyone from the mass media to politicians to Madison Avenue advertising executives are using them, it's all right; it's hip and hop. Companies like McDonald's, Pizza Hut, Kentucky Fried Chicken, Coca-Cola, and Pepsi try to woo black consumers by singing, dancing, and rapping the praises of their products.

From the mouths of whites, this hip-hop, rap, and ghetto slang seems to have less of an edge and takes on a wider acceptance because it has become deracialized.

Blacks are supposed to rejoice whenever our way of life becomes more mainstream. We seldom do. For we see in it a sanctioning that can only be granted by white society. In other words: If you're white, it's all right. If you're black, step back.

7. The White Take

In no way is this a one-sided issue. I spoke with whites who said there were things blacks often did and said that whites found racially offensive or insensitive. In fact, they were not at a loss for words or examples of ordinary interactions with blacks that left them feeling slighted.

I have done and said things to whites that I knew in my heart were wrong. Things that were based on assumptions, stereotypes, and others' opinions. I have reacted out of anger and frustration. I have done to whites what they have done to me.

Some of those interviewed did not want their names used, concerned that their perceptions, though valid, might be misconstrued as racist. Others were willing to be identified by name, believing their openness and honesty were steps toward a more fruitful dialogue between the races.

Here are themes that came up often:

The Race Thing

Many whites said they simply hated it when blacks turned innocuous things into a racial guilt trip.

"I remember once, a black friend complained about a white salesclerk in a grocery store putting his change on the counter instead of in his hand," said Sylvia Lewis, a white woman in her fifties who lives in Queens. "When I said, 'She did that to me, too,' he stormed off. Sometimes it has nothing to do with race, but I don't think he saw it that way."

Mrs. Lewis, who graduated with honors from Dartmouth College but had to take a job as a secretary, said she worked alongside blacks who had not graduated college but felt, nonetheless, that they were being discriminated against because she earned forty dollars more per week than they did.

"I also knew shorthand and could type sixty words per minute," she said, offering the same explanation she gave her black colleagues several years ago. "Even so, a few of the black secretaries were convinced they were being underpaid because they were black."

Repeatedly there were whites who described situations in their workplaces in which supervisors or coworkers were accused by blacks of race-based behavior. Although many whites experienced similar treatment by the same white collegues blacks had complained about, blacks were convinced race was an underlying factor.

When asked to list two or three little things that black people have said or done to her that she found offensive, Ms. Sherwood, the white postal worker who attended the focus group in Washington, wrote the following:

1. Told me that I am a good white person and have a good heart to interact with the black community.
2. Say that I'm different.

Of the things she's heard said by blacks about whites in general, Ms. Sherwood listed these three:

1. Can't cook.
2. Can't dance.
3. Acknowledge blacks in private but not in public, or acknowledge blacks when whites are alone but not when they are in the company of their white friends.

Ms. Sherwood said what bothered her most was the assumption by blacks that what made her a "good person" was the fact that she was willing to interact with blacks or have black friends.

"I think I'm a decent human being who tries to treat everyone the way I want them to treat me," said Ms. Sherwood. "Besides, there are a lot of white people who socialize and kid around or joke with blacks and aren't good, decent people. Race has nothing to do with who is and isn't a good person."

Several whites interviewed talked about how angry they get when blacks, especially those with celebrity status, play the race card.

I've often jokingly explained to white friends and colleagues that when blacks are born, we're dealt a hand of "race cards"—aces and kings—that can be used at our discretion to gain an edge or an advantage in a race-conscious world.

Some whites felt that Clarence Thomas threw down a race card during the 1991 Senate confirmation hearings on his appointment to the United States Supreme Court. Anita Hill, a black law professor, had accused Mr. Thomas of sexual harassment during the time she worked for him at EEOC. Thomas vehemently denied the allegations and called the Senate Judiciary Committee's investigation into them "a high-tech lynching" of "an uppity black." Many blacks agreed, saying the hearings smacked of an orchestrated setup in which a black woman was being used to destroy a black man.

An entire nation watched as every white male senator on the committee squirmed as Mr. Thomas played out his hand.

Mr. Thomas is now Justice Thomas of the Supreme Court, like it or not.

"I don't agree with Thomas's views on affirmative action, but I didn't think it was right what they were trying to do to him," Carol McRae said during the Washington focus group. "If the Senate could have made a case against Clarence Thomas based on his judicial credentials, you wouldn't have had an argument from me. But to use a ten-year-old alleged sexual harassment allegation, when all of us blacks know that some of the senators sitting in judgment of Thomas had far worse skeletons in their closets, was unfair."

Monique N., a white woman in her midforties who owns a vintage-clothing store in Santa Monica, California, heard arguments like the one put forth by blacks like Ms.

McRae and was thoroughly appalled. "Blacks wanted to excuse Thomas's behavior solely on the argument that... what... well, Ted Kennedy walked away from Chappaquiddick and a woman was killed because of his conduct. Or, the old guy... Wilbur Mills was caught frolicking with some dame in the Potomac River. So what! They weren't asking to be confirmed to a seat on the nation's highest court. Even worse was that blacks were saying that Anita Hill was being used by white feminists to bring down a black man. I didn't believe that for one minute."

It was one thing for Clarence Thomas to use the race card to his advantage, many whites told me, quite a different matter for O. J. Simpson to use the race card to, in the opinion of a white majority, "get away with murder!"

At a dinner party session of the focus group in Betteron, Maryland, whites wanted to know how many of the blacks present believed Mr. Simpson was guilty.

Of the eight blacks present, seven raised their hands.

The whites then wanted to know whether the blacks felt it was right for Mr. Simpson to be acquitted.

"It's not a question of right or wrong," said Diane Camper, a good friend. "The prosecution failed to prove its case beyond a reasonable doubt." She then borrowed a line from Johnnie Cochran, Simpson's lawyer: "If it doesn't fit, you must acquit."

Everyone laughed, but not for long.

Roy Schneider, the thirty-year-old son of Pauline Schneider, who'd injected the "big thing" of O. J. Simpson into my "little things" book forum, said he believed O. J. was innocent and didn't understand why whites "couldn't get over the O. J. thing."

"White men have been getting away with murder for years, and now one black man gets off and it's cast as a total miscarriage of justice," said Roy. "Whites are still threatening vengeance. They can't stand the fact that a brother had enough money and clout to manipulate the system. We won one."

"Why would you want to win that one?" asked Elissa, a young white woman. "Of all the ones to win, why that one?"

Dr. Manley, the black sociologist, is convinced that both blacks and whites are guilty of exploiting their race when it's convenient. "Whites use the race card too," he said. "To get entrée into exclusive clubs or schools or to get that job promotion. 'Our dads play golf together, so...' Whites have thrown down the race card to get out of sticky situations....'Do you know who my parents are?' What they didn't like was that O. J. had enough economic potential to neutralize the justice system in favor of his race. He could pay people off and thus was able to make race become the feature you had to deal with. Don't try me as a resident of Brentwood; don't try me as a celebrity; try me as a black man."

Dr. Manley said President Clinton also played the race card during the Whitewater and Lewinsky investigations. "He went in the opposite extreme," said Dr. Manley. "Clinton was saying, 'I want a personal life.'"

Whites may have also wondered why black Americans rallied around a black man who tried to live his life as a white man. O. J.'s second wife, Nicole Brown, was white; his girlfriend, Paula Barbieri, was white; he socialized almost exclusively with whites; his closest friends were white; his houseguest, Kato Kaelin, was white; and he lived in Brentwood, an exclusive white community in Los Angeles.

"Explain that to me," said Elissa.

They're right on this one, folks. Blacks publicly went to bat for Simpson. Some apparently suffered considerable grief for doing so. There were reports after Simpson's acquittal of white coworkers who stopped speaking to black colleagues who felt the verdict was just. There were reports of physical altercations between black and white college students. In integrated neighborhoods, it was black neighbor against white neighbor. In some cases, the damage done was irreparable.

And where was O. J.? Back on the golf course. Back with his white girlfriends. Back home in Brentwood, if only briefly. O. J. made one public appearance in South Central Los Angeles to thank black residents there for supporting him. Days went by before Simpson bothered to publicly acknowledge and thank the black Americans around the country who supported him.

But blacks still defend his actions, even those blacks who think he's guilty. I'm sure it's confusing to whites, because it's confusing to me. I'd like to say I know what's up with that, but I'm not sure I do.

It wasn't about class. O. J.'s biggest supporters were low-income blacks. It wasn't a gender thing. Black women were as supportive of O. J. as black men. In fact, after the verdict three of the black women on the jury were openly hostile toward another black female juror, who was leaning toward a conviction. Nor did support for O. J. break down along generational lines.

O. J. gave black Americans something to rub in white folks' faces. Something to get under and irritate their skin.

I don't for one minute believe some of the blacks who supported O. J. liked the brother. Privately blacks will tell you just the opposite: that they didn't like the brother or what he stood for. His arrest, imprisonment, and lengthy trial was payback for turning his back on the black community and trying to be white, many blacks believe.

At the same time, blacks cannot forget the countless numbers of white Americans who also got away with murder, and all the times our cries for equal justice fell on deaf ears. We cannot forget the times white police officers planted evidence on black suspects and got away with it. Some blacks have said they weren't so much pulling for O. J. as they were for Johnnie L. Cochran, his black attorney.

Johnnie was on the spot, and blacks wanted to see this charismatic black attorney beat the white man's system. His victory was seen as a victory for the race. That's what whites have failed to see in all of this. Not that O. J. got away with murder. But that Johnnie Cochran, a dark-skinned black man, who'd once been stopped in his Mercedes on suspicion of DWB, who is married to a black woman, who wasn't even an initial member of the Dream Team, got him off.

Curiosity Killed the Cat

It's human nature to be curious. We've all asked stupid questions about innocuous things. Yet whites who may be genuinely curious about black culture or style have often been accused of being racist simply by asking a question. No harm may be intended, but it's often felt, and many whites who have found themselves on the receiving end resent the implication.

"I remember when Vanessa Williams won the Miss America title," said Greg, a white advertising executive in San Diego. "I was telling some of my black friends how beautiful

I thought she was and you would have thought I'd insulted their mother. Their response was, 'Well, she looks like she's white, so I guess you would find her pretty.' I didn't get it. She is pretty, period."

Whites said they are often afraid to speak openly and freely around blacks, fearing that a misstep or a word taken out of context could place them in the awkward position of having to defend their words, actions, and race. "Better to remain silent," said one white male.

Elizabeth S., a white business consultant who, with her husband, Alan, owns a farm in the Catskills region of upstate New York, learned that lesson the hard way. "We had two blacks kids, a boy and a girl no more than ten years of age, staying with us one summer as part of the Fresh Air Fund program. I'd noticed that whenever the kids went swimming or took a bath, their skin became really ashy, and I wasn't sure what to do about it. So, one day I'm talking to a black colleague at my office and I happened to mention this, and the woman, whom I considered a friend, sucked her teeth and rolled her eyes and told me that I sounded so ignorant. 'Just put some lotion or vaseline on their skin after they bathe, for Christ's sake.' That's what she said and walked away."

Elizabeth said she took the woman's advice and the problem was solved, but she never said anything else to the woman about her two little summer guests.

A white male in Washington told me that a black coworker, whom he considered a friend, stopped speaking to him after he made an innocent remark about the black man's newborn son.

"He's so pale," I said to him. "And he said, 'Well yeah, so what are you trying to say?' He sounded so agitated that

I tried to explain that I was just making a comment. Afterwards, a black female friend explained that he probably took exception to what I'd said because he may have thought I was trying to infer that the baby couldn't have been his because he was dark-skinned and the baby was so light-complexioned. Once she said that I understood, but I hadn't intended to make any kind of inference. To this day my black colleague barely speaks to me."

Here's what my mother and grandmother told me on that subject.

Black babies are often extremely pale at birth. And I know some black folks who would prefer they stay that way. But as life would have it, they do get darker. If you want to determine how dark or light a black infant is going to be when they get older, look at the color of the skin around the baby's cuticles. It's an accurate barometer.

Many white infants, on the other hand, lack color in their hair at birth, which is often blond through their toddler years. The hair tends to darken as the children grow older. Unless it's in the genes.

"On the one hand, blacks accuse whites of being ignorant about them and the way they live, but when we make an attempt to find out more, they shut us out or snap at us for the least mistake; we're afraid to go there," said A. Douglas, an old army buddy of my brother Ronald. "People learn from their mistakes. If whites are too afraid to ask blacks the simplest question, we'll never know anything and resort to making assumptions about what may or may not be true."

A white male friend who lives in Los Angeles said he was reluctant to compliment black coworkers or acquaintances on certain things, be they achievements in the office or in life, because too often, he said, his well-intentioned observations were misinterpreted by the recipient.

"Like this one time, I was visiting the home of a black coworker and noticed that in his rec room were all these tennis trophies," he said. "I simply asked where did all those trophies come from. And he shot back: 'Well, where do you think they came from?' I didn't know what to say. He told me that he and his wife have been playing on the USTA weekend circuit for years. How was I to know that? This was the first time he said anything about it, and I'd never been in his home before that day."

So what made my white friend think that his black coworker mistook his inquiry as an insult?

"Because later that night, I overheard him telling a group of blacks there that I acted as if black people didn't know how to play tennis or weren't good enough to play on the USTA circuit. That's not what I meant at all."

Sometimes even the good intentions of white Americans become suspect to blacks.

In 1998 a white third-grade teacher was forced out of a school in the Bushwick section of Brooklyn by black parents because she'd read a book called *Nappy Hair* to the class. The teacher said she'd read the book to her class in an attempt toward multiculturalism.

Written by Carolivia Herron, the book is a story about a young black girl's appreciation of her kinky hair. The book has always been popular among blacks of various classes and regions who see in its tale a positive spin on what has long been viewed as a "negative" black trait. I bought a copy for my eleven-year-old niece, and together we read the story and marveled at the colorful illustrations.

But the black parents of students at Public School 75 took offense. They said the material was racially insensitive and belittled blacks. Others resented having a white woman read a book about blacks to young, impressionable black children. Ironically, several of the black parents admitted they had not read the book.

The teacher was so disturbed by the matter and the publicity surrounding it, she resigned from her teaching post—a victim of circumstance. I found myself empathizing with her.

What do black people want? she must have thought to herself. *They want white educators to stress diversity and multiculturalism. Then when we do, they complain. Damned if we do, damned if we don't.*

A white male executive named Jonathan, whom I met on a flight from Seattle to Miami in July 1999, said he worked for a prestigious international management consulting firm and, upon hearing I was writing a book on race relations, asked my opinion about the following incident that happened when he was working in Los Angeles.

"My firm had discovered that a lot of employers in the service sector were looking for young people with computer skills to work in the industry," he told me. "So, we decided to teach inner-city youths how to use computers and how to service computers, with the hope that they could then take that knowledge and, if not get a job at IBM, maybe find a job at Safeway food stores. When I told this to a black man who headed up a program for youths in East and South Central Los Angeles, he accused me of trying to prepare a new generation of young blacks for menial work that would have them servicing whites."

Jonathan said the thought had never crossed his mind. He, and his company, believed that giving young black youths access to computers and teaching them basic computer skills might benefit them in the future.

"But this guy said we weren't going to teach them enough to get high-paying jobs, only low-paying service jobs." Jonathan said his firm decided not to go ahead with the program. But that wasn't the only reason he wanted to share the story with me. A few years ago, he was transferred by his firm from Los Angeles to Seattle, Washington, and once again had been approached about starting a similar program for black youths in the Pacific Northwest. Jonathan passed.

"I was afraid black leaders in Seattle might feel the same way the blacks in Los Angeles felt about our intentions," he explained.

But You Call Each Other "Nigger"

"But you call each other 'nigger' all the time," a middle-aged white woman remarked during one of my focus groups in Washington. "I was riding the bus the other day, and a group of black teenagers kept referring to each other as 'nigger this' and 'nigger that.' If whites were to do that, blacks would get upset."

What's up with that?

I've been called a nigger by whites and I've been called 'nigger' by blacks. I don't like it either way. But it's one thing to be called 'nigger' by a member of the race, quite another to be referred to by that term by someone outside the race.

Nigger is one of America's most enduring epithets, a

racial slur that managed, through the ages, to defy the adage that "words can never hurt you." Whites knew—and still know—this. They know that by calling blacks "nigger" they strip us of our humanity and jar our collective memories, reminding us of days not so long ago when all of us were niggers in their eyes.

Behind closed doors we have referred to other members of the race as niggers. An older generation of blacks often used it to describe the "lowest of the low" among us. Never, ever, was the "*N* word" to be used in front of whites.

In the 1990s a hip-hop generation of black youths embraced the term, using it at will in rap music, film, and everyday conversation. In doing so they hoped to strip it of its painful racist legacy. To them *nigger* was more about attitude than race.

"Anybody can be a nigger," my nephew Deriek explained. "Color has nothing to do with it. It's used more to describe people who do stupid things... like robbing a bank without a mask. You hear somebody did that and, regardless of what color they may be, you say... What was that nigger thinking about?"

Niggers found their way into the American mainstream, from the movies of Spike Lee to the gangsta rap of Ice Cube.

With blacks commonly referring to each other as "nigger this" or "nigger that," inevitably whites felt they could take similar license in the word's usage. Some white youths, with a certain black sensibility, say they refer to each other as niggers, a practice that spawned a new word in the urban vernacular: *wigger,* which refers to white youths who "act black."

In a story that appeared in the *New York Times* in January 1993, Bob Guccione Jr., editor and publisher of *Spin,* the popular music magazine, told *Times* reporter Michel Marriott that whites are very reluctant to use *nigger* because it has "such an incredible weight of ugliness to it."

"In a sense, it empowers the black community in the white mainstream," said Mr. Guccione, who is white. "They can use a very powerful word like a passkey, and whites dare not, or should not, use it."

No matter how close whites may be to black Americans, the word *nigger,* when uttered from their mouths, has not lost its power to wound and inflict pain.

"Quentin Tarantino saying 'nigger' in *Pulp Fiction,* is different to me than, say, a white cop calling a black man a nigger," Deriek explained. "That's what I think white folks don't understand. Even though it's been deracialized, it doesn't mean every Tom, Dick, and Sue can say it and get away with it."

With due respect to my nephew, I took offense at Tarantino's constant use of the *N* word (my linguistic preference for that term) in his 1996 movie.

"If blacks don't want whites to use the word, why are they putting it out there in the mainstream popular culture?" a white female colleague, who is also a close friend, asked me one evening over dinner at a Manhattan restaurant. The conversation took place shortly after the movie opened and was between friends, thus, I will not use her name. "Blacks refer to each other as niggers all the time, so whites feel . . . well, maybe it's all right to use the word in certain cases."

Using that logic, I guess, one might surmise that as

long as parents call their son a bastard, it's all right for a stranger to call their son a bastard. I DON'T THINK SO!

Still, it must be confusing to whites. A white man who calls a black coworker a nigger might be suspended from the job, or worse. But a black man can call another black man a nigger and not even be called on the carpet for it. F. Lee Bailey, the white trial lawyer who was a member of O. J. Simpson's defense team, considerably damaged the prosecution's case by planting a seed in the jury's mind that Mark Furhman, the white Los Angeles police officer who testified in the case, was a racist. Why? Because the detective routinely referred to blacks as "niggers." But Mike Tyson is all but called a nigger by one of his white attorneys who asks for and receives a more lenient sentence because of it. *Nigger* is the racial epithet most often cited by law enforcement officers as evidence that a crime against a black American may fall under the Hate Crimes Act.

Blacks have used the term to their advantage, and whites know it. Which is why we all should just leave *nigger* alone.

Light, Bright, Damn Near White

A white male acquaintance asked me about the "brown bag test." I'd rather he not have gone there.

He was referring to a rather painful chapter in black history, one of our so-called dirty little secrets.

During the 1930s and 1940s, light-complexioned blacks, many of whom were mulatto descendants of slave masters, felt their paler brown skins made them "better" than darker-skinned blacks. In order to maintain the few privileges granted lighter-complexioned Negroes, some black social clubs refused to admit blacks whose skin was darker than a brown paper bag.

"But doesn't that show that black people can be prejudiced amongst themselves?" my white acquaintance wanted to know. "My only point is that a lot of times blacks act as if they don't have a prejudiced bone in their bodies. But I've heard some of my black friends and coworkers say things that I frankly find prejudiced and offensive."

He didn't have to go into specifics.

Yes, we do it. Blacks discriminate against each other, more than I care to admit. There are blacks who will only date blacks who are, as the saying goes: "light, bright, and damn near white." Black men who date only black women with long or straight or curly textured hair.

Psychologists say it is part of the self-hatred that's been bred in us. I buy that, sometimes. But this whole thing about who's black and who's not black has gone from the sublime to the ridiculous.

Tiger Woods was rebuked and scorned by many blacks when the pro golfer let it be known that he was of mixed heritage: His father is black American, his mother is from Thailand.

"Fine. If that's what he is, that's his heritage," said Joanne Williams, my sister-in-law. "But let me say this about that. If Tiger Woods robbed a bank, they wouldn't be saying they're looking for a suspect of mixed heritage. They'd say they were looking for a BLACK MAN!"

The Jim Crow definition that a "single drop" of black blood made a person black didn't lose credence because blacks objected, but because whites did. Why? Because it meant that a great many Southern whites weren't so sure they were "racially pure."

But that's beside the point raised by my white male friend. We'd once dated, until I let it be known that as far as I was concerned a drink or dinner and a kiss good night were as far as I was ever going to go in our relationship.

"Why?" he asked on one long walk home back in the mid 1970s.

"Because I don't sleep with white men," I told him, never thinking twice about how that may have sounded.

"Then you're prejudiced against whites," he responded.

"Me, *prejudiced*?" We managed to laugh and have remained friends to this day. Years later I jokingly told him that he was one of my few regrets in life, to which he suggested I could still make amends.

I know what often happens to blacks who date or marry whites. They're ostracized, shunned, labeled traitors. I wasn't about to chance such isolation. Many blacks would sooner forgive Mike Tyson for sexually molesting a black woman than they would another black for marrying a white.

"My only point is that blacks act as though whites are the only people who harbor prejudice," my friend noted, in concluding his train of thought. "Blacks can say they don't want their son or daughter to marry a white person, and whites are supposed to understand. But let a white person say something like that."

And blacks are prepared to bring in the National Association for the Advancement of Colored People.

I have relatives and close black friends who are, or have been, married to whites. I've dated white men. But whenever I see a black man with a white woman, I get this queasy feeling inside.

I'd like to say I've gotten over it, but I'd be lying if I did. Each time I see the mixed coupling, I suffer a tinge of indignation. Perhaps because I see their liaison as a rejection of black women. What hurts even more is when prominent blacks—celebrities, athletes, politicians, business entrepreneurs—marry outside the race.

Many of us remember the days when miscegenation was a crime in certain states, punishable by imprisonment. We heard stories of black men being lynched for nothing more than looking at a white woman. There were accounts of black men who deigned to sleep with a white woman, only to have their mutually agreed-upon sexual encounter declared rape.

Whites see our actions as racist. In some ways they may be. But they are based on the singular belief that interracial marriage is another form of black self-hatred. Blacks who say they cannot find members of their own race to marry are playing into the white man's perception of beauty and his concept of black inferiority. Beyond that, many blacks wonder whether whites who marry blacks who have accomplished something in their lives are doing so out of love or because of the status they acquire in the marriage.

There is a fundamental belief among blacks that white women only go after black men who "have something."

"When you see all these black athletes and black entertainers marrying white women, you do wonder whether it's for love," said Jackie DeShields, who works with my sister at the post office in Washington. "You have to wonder if these white women would be marrying these brothers if they worked for the sanitation department or McDonald's. That's what bothers me. They tend to go out with the brothers with money.

"And why is it that black men are content with white women who aren't that attractive?" she added. "I've seen these good-looking brothers with these plain-Jane white women."

Jungle fever, blacks call it. Spike Lee even did a movie about it. Of course, it's offensive to whites.

"My sister-in-law is white, and I know she loves my brother," said Leslie Lewis, who I will forever think of as the little girl who went to Nativity Catholic school with my niece Antoinette, even though she is now a mother of two small children. "I see people giving them the eye when they walk down the street and I resent it 'cause she's not that type, and she's attractive."

Love can transcend race. Pauline Schneider, my black lawyer friend, illustrated that point when she shared with me the story of her brother.

While in high school in the 1950s, her brother, Sydney Boykin, had fallen in love with a white student named Linda and she with him. Back then such things weren't tolerated, so they hid their feelings. The two graduated, went to different colleges, got jobs, and eventually met and married people "of their own kind."

A few years ago, the two bumped into each other. Both were now divorced from their spouses. Now, nearly half a century later, they began dating. When I met them in Betterton that Memorial Day weekend of the focus group meetings, the two were engaged to be married.

No matter how you feel about the races, that's a lovely story. And you can't tell me that these two people don't love each other. Over dinner that evening Linda talked about the racism she has seen as a white woman engaged to a black man.

"We were waiting in line at a store and the saleswoman obviously assumed I wasn't with Sydney," she said of her fiancé. "He was patiently waiting for service, and this woman was completely ignoring him and taking care of every white person in sight. Finally, I just spoke up. I said, 'We've been standing here for several minutes waiting for service, and you've waited on people who came in after we did.' I could tell then that she realized we were a couple. She denied it, but I knew what was going on. I left the item on the counter, and we left."

Whiteness carries a certain status. Whites know it; so do blacks. Many whites feel that deep down inside what black people really want is to be white. That if left to our own devices, blacks would switch places with them in a New York minute. That's why blacks who achieve a certain social or economic status in life marry white. Doing so enhances their station in life, they assume. The question is, in whose eyes?

Whites have a point here. Blacks have to come to terms with their own color hang-ups if we're going to hold whites accountable for theirs.

The Macho Factor

A white man in his late forties talked about black men who usurped space on subway cars by sitting wide-legged, causing others to be sardined on their seats. "They sit there defying you to ask them to move over, as if their manhood is at stake," he said.

Another white male said he resented inferences often made by black men that they were somehow more masculine or macho than whites.

Hair may be a primary irritant between black and white women, but when it comes to men, it's all about machismo.

Howard A., a white male friend, told me the following story.

He was walking hand in hand with a black female companion on the downtown streets of Columbus, Ohio, when a black man, a complete stranger, accosted the two.

"He mumbled something like: 'Oh, you're into that, huh, sister?' I was about to confront the guy, but she pulled me away, saying, 'Just let it go,'" recalled Howard. "That made me angry, because I thought she thought I couldn't take this guy or something. It may not have been that way, but to this day, that whole incident bothered me."

I understood his anger and disappointment. I've had black men give me a funny look when I'm with white male colleagues or friends in public places. I know they're trying to figure out the nature of the relationship: friend or colleague. God forbid, the white guy next to me might actually be my lover!

Over the years, I've been asked by black men—friends, colleagues, relatives—whether I find white men attractive. Well, here it is: "It depends on the white man."

I would be lying if I said that in my forty-something years on this earth, I've never been attracted to a white man. I have. Though you wouldn't have gotten me to admit to such a thing in the past.

Besides, you know what they say about them—the "them" being white men. That they're not well-endowed, if you know what I mean.

A white male postal worker in Washington asked me whether I was going to talk about the "size" issue.

I knew exactly what he was referring to, and I really didn't want to go there.

"I'm writing a PG book," I said sheepishly.

"Yeah, I know, but it's one of those little things—no pun intended—that drives us white males up the wall."

He was talking about penis size. And he was right.

I couldn't really deal with the things that irritate and annoy whites if I didn't talk about the size issue.

For years white men have had to deal with the stereotype that they are not as well-endowed sexually as black men. For years black women have been talked out of dating or marrying white men because we have been lead to believe—by blacks in general—that they lacked the "adequate tools" to satisfy us in bed.

Most of us accepted this as fact. How we as a people knew this very private fact was never fully explained. In a segregated nation, with separate public accommodations, blacks seldom were allowed to get close enough to white men to find how they were endowed.

Nevertheless, the perception became accepted reality.

"I refused to get undressed around the black guys in my school because I always felt they were looking at me and measuring me against them," said Kevin, a white lawyer now in his late twenties, who continues the practice to this day, when he goes to his integrated New York health club. "I always resented the assumption. I always felt it was as wrong for black men to say that about us as it was for us to assert that they were these overendowed beasts with tails."

Such a sensitive subject would never find its way into one of the president's forums on race. Imagine how that might go over in prime time. Yet it has long been a source of private conversations and whispers in intrarace group settings. And it continues to fuel stereotypes and bruise emotions. My black male friends say the stereotype persists because it gives black men a perceived advantage over white men.

"It's the one thing we can use against them," said Raymond Samuels of Washington, "to get them all worked up. That's why a lot of black men go after white women. Because it leaves white men thinking that 'once you've tried black, you won't go back.'"

A counterpoint to his point was provided by a white male friend: "Once you've tried white, you'll be all right."

"The whole size issue is a farce," said H. T. Starr. "I read a study that said when it comes to size, French men are more endowed than either black or white American males. But it's something that's been whispered about for centuries," he said.

So. We're supposed to believe that not only do white men come up short—again pardon the pun—in terms of virility, they also can't jump!

Every day, the racial dynamic of machismo is played out on the basketball courts of America.

David Neidorf—a white man in his thirties who grew up in an integrated neighborhood in Los Angeles, attended integrated schools, and now teaches at an integrated school in West Los Angeles—recalled playing a pickup basketball game with a group of young black males.

"I was being elbowed and kneed every time I touched the ball," he recalled. "I knew it wasn't just my game and style of play that was being challenged on that playground. It was also my manhood. Finally I'd had enough and I got in one of the black guy's faces and asked him if he wanted a piece of me. He looked at me and smiled and said, 'Chill out, bro. It's just a game.' After that, all the pushing and shoving stopped. But it wouldn't have if I hadn't had the guts to confront that one guy."

Some white men, including coworkers of mine, said that most black men act as though they can beat, pugilistically speaking, any white man on the face of the earth.

It's true. I had a black friend get his butt kicked by a white guy in a bar in upstate New York during the 1980s because he assumed he, my black friend, could whip this guy. A lesson learned the hard way.

John, a twenty-something white male, recently married, and also in Betterton for the weekend, said he would be more reluctant to challenge a black male than a white male.

"In part because the perception of black males is that they're tough," said John, a bespectacled man of medium build. "But then I'd regret backing down, because my manhood couldn't take it.

"And that's what leads to confrontation," said Larry, his father, who's noticed another pattern among black youths who see him driving a Mercedes.

"They take their time passing in front of the car," said Larry.

The blacks present took strong exception, wondering aloud if the reaction was not just that of youthful machismo, rather than race-based.

"I'm just saying that black kids do it to me all the

time," he explained. "They see me in my car at an intersection or traffic light, and even though I may have the right-of-way, they just slow down to a snail's pace and glare at me, almost daring me to honk or show any outward displays of emotion."

I had to laugh. Although I wanted to agree with the blacks there, I conceded that Larry probably had a point, because I have been known to dare whites driving luxury cars to honk or to try to intimidate me into moving any faster than normal.

"Go ahead and the car's mine and that's just for starters," I yelled at a white woman in a Range Rover who seemed bent on turning the corner before I could cross the street.

And I didn't stop there. I snarled at her, slowed to a snail's pace, and mumbled, "Act like they own the world." The "they" meaning white people, not people who drive Range Rovers.

While young men of all races and creeds may have responded similarly to Larry and his Mercedes, seeing him as a symbol of the establishment or the privileged class, many blacks will see the additional element of race involved. Thus, he becomes part of the white establishment or the white bourgeoisie.

Despite the protestations of the blacks present, Larry felt his perception of these brief encounters was reality. He was right. If he even thought about getting out of his car to confront those black youths or began honking his horn to move them along, a little thing would have erupted into a big thing and who knows what.

"When you kept your cool, it probably caught those

kids by surprise," I told Larry. "And that probably got under their skin more than if you had taken the bait!"

A "Black Thing"

"It's a black thing, you wouldn't understand" was the slogan on a popular T-shirt a few years back. Many whites took offense because the slogan implied that whites lacked a level of understanding about African Americans.

"It's not important whether it was intended to hurt, but how it's perceived," said Mr. Weisman, the peer counselor on race relations for the Anti-Defamation League. Mr. Weisman asserted that if the adjective were changed to *white* or *male* or *female,* it wouldn't be difficult to understand why those not in that group might take offense.

He's right. But some blacks embraced the saying simply because they knew it got under white folks' skin. It suggested a degree of ignorance on the part of whites about black life. No need to explain it to you, 'cause you wouldn't understand. It's the height of condescension. Blacks knew that and used it to their advantage. Whites played into our hands on that one by showing indignation. Worse still, when the phrase was turned around—"It's a white thing, you wouldn't understand"—the black retort was "Yeah, probably not." That, too, bothered whites.

"'Cause they know when we say that, we're implying it's something stupid that only whites would do," explained my nephew Mark, who like most young blacks noted that black *thang* was used to give the slogan more of an edge. "Like getting their thrills by skydiving or climbing some snowcapped mountain. I mean, not many black folks I know enter those hot-dog-eating contests at Coney Island.

That's white folks' foolishness. A white thang, we wouldn't understand."

Maybe whites are more daring than blacks when it comes to tempting fate. Maybe it comes part and parcel with white privilege, this feeling of immortality.

Being Hopelessly Uncool

The idea that whites don't have rhythm, can't dance, or just aren't as hip as blacks has long been a source of irritation to many whites, especially those of a younger generation.

"It's as if blacks have a monopoly on hipness," said Susan, a Forest Hills, New York, teenager. What hurts even more, a few whites indicated, are the suggestions that whites who do appear to be on the cutting edge are "acting black."

"A white male friend of mine wore an Afrocentric shirt to school, and some of the black kids said, 'Why are you trying to be black?'" said Rachel Weiss, a student at South Shore High School in Brooklyn, New York. Ms. Weiss, who also counsels other students for the Anti-Defamation League, said she responded with a question of her own: "Why can't he wear it?"

As if blacks own red, black, and green.

David Smith, a Jewish man married to a black woman and the editor at the *New York Times* whose idea it was for me to write the story on the "little things," asked me one day whether I'd ever come across a white person I considered hip or cool.

"Yes," I replied. Knowing he'd want a few examples, I cited James Dean and the Righteous Brothers, the 60s white duo whose soulful rendition of "You've Lost That Loving Feeling" led to their being dubbed blue-eyed soul brothers. I mentioned Michael Bolton, the 90s version of the Righteous Brothers (minus one), and Robin Williams, the actor and comedian. I pointed to President Clinton and Elvis Presley. Yeah Elvis's name was on my list of hip white folks, even though blacks have never forgiven the King for his comment that the only thing a black man could do for him was shine his shoes. The question was about hipness, not intelligence.

So why did David want to know?

Being an editor, he had a story in mind. He believes an evolution has taken place in American culture over the past five decades, one in which being cool meant being black.

"If you were to ask anyone, black or white, what I asked you, I'd bet few would have a white person at the top of their coolest people list," he explained. "It'd be interesting to find out."

I have this white male friend named Paul. He's hip—for a white guy. I say that to say this: Blacks don't expect whites to necessarily be hip. They're hopelessly uncool. Whites who show any sense of style—in fashion or manner—are often accused of trying to be black. If they speak in the vernacular of inner-city black youths, they're trying to be black. If they can dance—and yes, there *are* whites who can dance—they "must have some black blood in them."

"Or, it's because we've taken dance lessons," my white friend Jill Gerston-Newman said with a chuckle. Jill, who often parties with black friends, admits to being self-conscious about dancing around them.

Two white female colleagues at the *Times,* who could do the electric slide with the best of them, told, for non-attribution, that even when it's obvious that there are whites who have rhythm, blacks won't give them the credit due.

"Remember when *Saturday Night Fever* was the rage?" asked one of the women, who said she learned the slide dance while attending a black wedding. "None of my black friends would say—to me, anyway—that John Travolta could dance. Nooo. All of his steps were choreographed, they said.

"And remember *American Bandstand,*" she went on, referring to the popular 1960s teen dance show hosted by Dick Clark. Both she and I were old enough to remember. "A lot of blacks watched that show and enjoyed it. But I remember talking to a black kid who went to my high school in Queens, and he said that the white kids on *Bandstand* knew how to dance because blacks had taught them. When I tried to argue with him, telling him that the kids on *Bandstand* came from all over the Philadelphia area, so how could that be, he said they were going to black clubs. Then he said that even though the *Bandstand* kids could dance, they couldn't outdance black kids."

The idea that whites can't be naturally cool appears to be accepted truth in the world of sports. White ballplayers who are fleet-footed or who can dunk or play with an aggressive school-yard style have been told that they play like they're black.

"You must have some black blood in you," Kevin Little, a white sprinter has been told.

Black ballplayers, whether on the basketball court or the football field, "play with style." Whites...well, they have a "nice style to their game." Black athletes even have the hipper nicknames. Compare black—Air Jordan, Prime Time, Chocolate Thunder, Magic, and Brown Sugar—to white—Bird, Big Red, the Snake, and Pistol Pete.

Mike Schmidt, the white Hall of Fame third baseman for the Philadelphia Phillies, was once accused by some of his teammates of acting black, because he walked with a slow, purposeful swagger popular among black male athletes. Schmidt, to his credit, said they were probably right, but he attributed his gait to the fact that he'd been around black men most of his adult life and maybe some of their strut had rubbed off on him. That ended that little controversy.

Coolness is something for blacks, as a people, to be proud of. As such, we're not about to relinquish our hold on it anytime soon.

TALKING OUT LOUD!

Mrs. Lewis, the white woman from Queens, says she is not one for generalities, but . . .

"Blacks talk out loud in movies, and that really annoys me," she said. "What they do is talk back to the screen. They'll just provide the dialogue or react to something they've seen."

Mrs. Lewis, who asked to be identified by her maiden name, said she believed that blacks who behaved that way were "flouting the rules of white etiquette."

A. Nadler of Queens wrote a letter to the editor of the *Times* to tell me he agreed with Mrs. Lewis.

"Blacks have ruined movie experiences for me and deprived me of subway seats. I, too, am man enough to get on with my life. But in scale of indignities, upsetting other paying passengers or a theater of paying patrons seems a somewhat greater offense than asking someone if they sunburn."

Mr. McCaffrey, the white actor from Los Angeles, said the talking out loud isn't what's made him think twice about going to movies that draw large, black audiences.

"I find that whenever I go to see a black film, especially a comedy, if I laugh at something I find funny, I get these snide comments from other blacks," Mr. McCaffrey told me. "In the movie *Hollywood Shuffle* there's a scene where a black actor keeps putting hair spray on his geri curl. Every time he shook his head, the spray would splatter everybody. Everyone in the movie theater was dying laughing and so was I. Then I hear this guy behind me say, 'What you laughing at?' I turn around and I see the black guy looking at me, gesturing with his hands, like, what... you want a piece of me? I just turned and looked back at the screen."

In a letter to the editor I received after my story appeared in the *Times,* a white woman wrote to say that she gets annoyed when blacks take the liberty of ad-libbing dialogue in a movie. The woman said she felt that blacks who did this were trying to show off.

"They may think they're cute or funny, but they're not," the woman wrote. "Some of what they're saying is vulgar and uncouth. Somebody needs to tell you this."

We are a vocal people. I must confess that when I just want to sit back, relax, and enjoy a movie, I often go to theaters in predominantly white areas. And I'm not the only black American who does this, which has probably freed me to out myself here.

Older blacks say talking out loud in the movies started

with the release of *Birth of a Nation*. My grandfather John Quincy Adams told me that blacks were so incensed at the way they were portrayed in that film, they began shouting obscenities at the screen. Others shouted sarcastic retorts during certain scenes, which made the other blacks in the theater laugh and the whites suck their teeth. At that point blacks knew they had hit a white nerve and refused to let it go. And the rest, you might say, is history.

They Don't Look Their Age

"Black don't crack."

It's one of those in-jokes shared among black Americans, because we tend not to look our age.

Some have suggested the melanin in our skin helps reduces the exposure to sun and, thus, we tend not to age as fast as white Americans, who have less melanin. Another popular theory making the rounds is that black skin tends to become ashen when dry. To keep our skin clear and smooth, we use lotion or vaseline. Some say this constant use of moisturizers helps reduce aging.

Whatever the reason—and again we can't generalize—the feeling is that white people's fair skin cracks faster than blacks, making whites look older than their age.

"I've had blacks say that to my face: 'You know white people don't look their age,'" said Jack, a white male who lives in Silver Spring, Maryland, and is approaching fifty. He believes his receding hairline and horn-rims make him appear older than he is. "A black colleague told me to my face that whites are always trying to be so damn young. It hurt because I felt the person was talking about me, even though she insisted she wasn't."

Jack said he preferred not to be identified by name not

necessarily because he feared repercussions for his points of view but "because I don't want people to know I'm sensitive about my age and my looks," he said with a shy smile.

Jack's wife, Marian, said the age thing bothered her, too.

"No, I've never been told that by a black person directly to my face, but I've heard it said of whites," said Marian. "I'm not saying that it isn't true. I really do believe that most blacks tend not to look their age. But that doesn't mean whites all look older than they actually are."

Blacks are convinced that whites are so sensitive about the subject of aging, they lie about their ages. "They want to be so young," said Marquette Miller, a black woman who looks to be no more than thirty, even though she is in her forties. "I think that's why they look older, because they keep pushing back their ages. Whites who are actually in their late forties or fifties are trying to pass for people in their thirties and forties. When you do that, it makes you look older because you are older."

Mrs. Lewis said blacks do not exaggerate when they say that they tend to look younger than most whites.

"I don't know the reasons why," she said with a shrug. "But that's not the issue. It's just not nice to go around telling people: 'Oh, you look much older than thirty-this, or forty-this or fifty-this. And to generalize about all whites like that. Well, it's not right."

After I'd finished writing my story, "The Little Things," for the *Times*, Mr. Smith, my editor, kept prodding me for other examples that might be cited in the section on "The White Take." By then I was too creatively drained to think.

It wasn't until I went home that Friday night and was watching a New York Knicks game on television that the age thing came to me.

Jeff Van Gundy, the white coach of the Knicks, is a man in his early thirties. He looks older. Maybe it's his balding hairline or the bags under his eyes or his pale gaunt look. Maybe it's the cumulative toll exacted on a man who has to handle egotistical multimillionaires for a living.

Whatever the reasons, watching him on television that night, I was reminded of the stereotype about whites not looking their age.

"How could you possibly forget that one?" one of my black colleagues asked. "There isn't a black person in America that doesn't believe that one. We know it's the truth, and so do they. I like to think it's one of God's little equalizers."

Aren't You Trying to Look White?

Many whites say they don't understand why blacks say they are proud of their heritage and Negroid features, yet do everything they can to look white.

"Black women relax their hair to make it straighter," said Caitlin M., a graduate film student at the University of Southern California. It was a Saturday afternoon in April 1998 and Ms. M., which is how she wanted to be identified, had dropped by a friend's apartment and found me there talking about the book. After a little prodding on my part. Caitlin opened up.

"I mean, blacks bleach their skin to make it lighter. They dye their hair blond and wear colored contact lenses. And now, all these black entertainers are having their noses narrowed and making their lips thinner. But if I were to say blacks are trying to look white, they would take offense."

It is probably confusing to whites.

Sometimes I can't figure that one out, either. For a while there, every young black starlet in Hollywood was going blond, Oprah Winfrey turned her brown eyes green, thanks to contact lenses, and Michael Jackson has grown up to be a successful white man.

Are these blacks trying to look white? Maybe they're simply making a fashion statement (although I doubt that in Mr. Jackson's case).

Imitation is a form of flattery, but it has its limits.

Just because whites go out and get a suntan doesn't mean they want to be black. Similarly, just because blacks straighten their hair or wear tinted contacts doesn't mean they want to look like whites in every way, shape, and form.

Thin Lips, Flat Butts

My mother always told her children that it was wrong to make fun of other people's physical appearance.

I took her advice and tried to keep my opinions of others' looks to myself. But there have been times in my life when I couldn't resist the temptation of making snide comments about whites. For me, and millions of other blacks, our actions were a way to strike back at a people who routinely denigrate black traits.

"They have no behinds," said Kathy Williams. "Their lips are thin. They're pale. But they act like they're the most beautiful people walking the face of the earth."

A white woman I've known for many years told me she's always been extremely sensitive about her flat behind, and she didn't like it one bit when one of her black female coworkers said her "butt was as flat as a pancake."

"I've always had a flat behind, and the older I get, the more it sags," she said. "But I'm not about to inject silicone into my behind to make it fuller.

Funny, how life is.

Many blacks say they have to be careful in choosing clothing, because most white designers do not make clothes to fit black people's fuller rear ends. On us, dresses designed for white women's flatter behinds tend to be higher in the back than the front.

Tennis dresses weren't designed with the full-figured black woman in mind. A generation of black female tennis stars played in tennis whites that rode their behinds. With all those creative designer minds in Paris, Milan, and New York, one would have thought that someone could have come up with a simple white dress that would have conformed to Zina Garrison Jackson's butt.

White women, on the other hand, complain that as they grow older, their behinds tend to sag, so if they have even the slightest pot of a stomach, their dresses rise in the front and sag in the back.

At the same time, I've heard white men complain that because of their flat behinds, as they grow older, they are often forced to wear suspenders to keep their pants from sagging.

Blacks think their fuller bodies age more gracefully than whites' leaner builds. It's another little dig that annoys whites.

There are whites who defy generalizations. They're blessed with good genes.

Also. Some of how you look has to do with how you live. I've known blacks who have abused their bodies by abusing drugs or alcohol and by living a fast life, who look just as old as, and older than, whites half their age.

Affirmative Action Baby

One of the most contentious issues between blacks and whites is affirmative action. A growing number of white Americans blame affirmative policies for creating divisions among the races and for all that is wrong with America itself.

"It's reverse discrimination," said one of my white female coworkers. "Yes, it was wrong for this country to discriminate against blacks. But you don't rectify past wrongs by discriminating against whites. That's what affirmative action does."

Blacks see it differently, of course.

"It is the root of most, if not all, of the racial tension between blacks and whites," said Lydia Pitts, a friend and corporate attorney, who fits the composite of "an affirmative action baby." Raised in Chattanooga, Tennessee, educated at Howard University, Ms. Pitts rose through the ranks of a Fortune 500 company to a vice presidency, one of the few black women currently holding that post at her Manhattan-based international firm. "Whites view it as a form of reverse discrimination. Blacks view it as a necessary evil, if you will, but we also know that it often undermines our achievement because it allows whites to dismiss any advancement we as a people have made by attributing it not to our talents, intellect, or hard work but to affirmative action."

Wherever I traveled in researching my book, the issue of affirmative action came up.

"But that's not a little thing," protested one black man at a focus group in Washington. "We'll never settle that issue. I thought we were going to discuss the little things."

Many whites I interviewed shared his sentiment. The minute the issue of affirmative action was raised, they began to squirm.

"That's a no-win discussion," said John, a white man who lives in Boston.

Nevertheless, many whites told me privately, that the one issue that riles them more than any other regarding race is affirmative action. Blacks know this. And we respond, in kind, because we know they blame us for something that wasn't our doing.

A little history here: White folks have been blaming us for everything from busing to racial quotas. We didn't ask for busing or affirmative action. Those policies are the doings of—in the words of Andrew Young—"smart-ass white boys." They were the most expedient remedies to past wrongs; the paths of least resistance taken by white lawmakers compelled by white judges. (Remember, blacks weren't in Congress in sufficient numbers to swing votes when those laws and policies were passed.) To say we made you do it is exaggeration by any stretch of the imagination.

Affirmative action was America's attempt at spreading the wealth, in this case, jobs. It was black folks' long overdue "forty acres and a mule," the unfulfilled proposal to compensate former slaves for their servitude.

Whenever I'm around whites who are ranting and

commiserating about affirmative action, I try to cure their selective amnesia with this historical antidote.

In its 1954 *Brown* v *Board of Education* decision, the Supreme Court declared separate-but-equal schools unconstitutional. It was then left up to the states to figure out a remedy. Now, it would have been quite costly to build new schools in the black communities across America and to stock those schools with books and supplies equal—be it in quantity or quality—to those in white schools. Moreover, the Court found that because blacks were not afforded the same quality of education as whites, black teachers were often lacking in their educational skills. So now, we're talking about additional education and training for black teachers. Pretty soon, we're talking billions of dollars.

The most expedient thing to do was to put little black children on buses and send them to crosstown white schools. Does anyone out there really think that any parent, black or white, would have purposely chosen to put five- and twelve-year-old children on buses and ship them halfway across town to get an education? Whites came up with this brilliant idea, but from Boston to Birmingham, blacks were being blamed. Like we held guns to their heads and made them do it.

Affirmative action is another example of what happens when you have what I like to call one-white-boy-too-many in the room.

The 1964 Civil Rights Act bars discrimination on the basis of race, color, creed, gender, disability, and religion. The act also designates which groups would be protected under the last—another bone of contention among blacks (but we'll deal with that later!). So that the law would carry weight, Congress agreed that those companies, educational institutions, or individuals who violated the act could lose federal funding and/or contracts with the federal government.

Interpreted literally, the law promotes the use of affirmative recruitment, and outreach and training programs to enhance employment opportunities for those who have been denied consideration in the past.

However, problems arose with the implementation of the law. Many white employers claimed the law was ambiguous. Did it mean employers were expected to just go out and hire blacks, regardless of their qualifications? Would one black do? What if you just couldn't find a "qualified" black, even though you tried? Should or would a company be penalized for trying? Once blacks or "minorities" were hired, were they supposed to be given preference in promotions as well? Could a company be fined for failing to promote blacks in equal numbers to whites? White colleges and universities were just as perplexed.

They, too, didn't mind opening their doors to minority students, but come on... fair is fair. Were colleges and universities supposed to admit black students regardless of their SAT scores and high school grade averages? Wouldn't that be unfair to white students? Much like employers, the colleges weren't sure whether admitting a few blacks would satisfy the government. They couldn't be expected to open their Ivy gates to every black who applied, could they? Colleges are also employers. Did they mean...? Oh dear! That meant colleges and universities would probably have to increase the number of minority professors on their faculty.

The law failed to specifically address many of these questions, leaving it up to the employers and the institutions to decide.

Some companies responded by hiring one or two token blacks. Universities admitted one or two blacks and patted themselves on the back for having complied with the law.

Once in these institutions, blacks soon became pariahs—avoided, shunned, denied access to inner circles, stuck at the bottom rung of the ladder. Black students complained they were graded harder than white students, seldom called on in class, and generally made to feel unwelcome.

Blacks felt like the hero of *The Spook Who Sat by the Door,* the 1969 novel about the only black man with the CIA. He is assigned a desk near a front door in the agency's office, far enough away from whites but visible enough for all to see. He is expected to be a racial showpiece, but he secretly organizes urban black guerrillas.

Laws were made with built-in loopholes, and white institutions found them in the affirmative action legislation. Concerned about further legal action, the federal government took steps to force companies to comply with the act, by establishing numerical hiring standards, called racial quotas, and timetables in which to meet those standards. The government set a goal of 15 to 20 percent—numbers equal to the percentage of blacks in America—as an accepted standard to comply with the law.

What riles so many blacks is that the government, whether by intent or design, also gave white Americans a face-saving tool to comply with the law by expanding the category of minorities to include women. When it came to choosing a "minority" candidate, many employers simply hired white women, instead of blacks, and thus were in full compliance with the law.

Statistics show that the biggest beneficiaries of affirmative action have been, and still are, white women, not black Americans. Once Aleuts and Eskimos were added to the class, blacks feared whites would extend preference to members of these groups instead of black Americans.

Few companies lost contracts with the federal government for failing to comply with the law. Few colleges or universities were ever fined for violating the law. Some schools, like Bob Jones University, refused to accept federal funding rather than comply with the law.

Others complied, but said it wasn't easy. Many schools and employers said it was difficult finding "qualified" blacks to meet their standards of admission or employment. As such, they wanted us to believe they had to lower standards or be penalized by the government. So the government let these poor little things off the hook by telling them that as long as they put forth a "good-faith effort" to comply with the law, they would not be fined. Now wasn't that special.

Define *good faith*? Sending recruiters to job fairs at black colleges and universities? Placing advertisements in *Ebony* and *Jet* magazines? Asking around town if anyone knew "a few good blacks" white companies could hire? Put out an "all blacks bulletin"? All a company or university president had to say was that they put forth a "good-faith effort" and it was end-of-story.

"A lot of the white companies didn't even want to do the 'dirty work,' so they hired blacks and appointed them as heads of their equal opportunity departments," said Ria

Griffin of Houston. "Blacks weren't in policy-making positions. We weren't meeting with the chief executive officers of these companies. We were put in charge of the other blacks."

The old H.N.I.C.: Head Nigger in Charge. Yes, my using that term is appropriate because it is in context.

Nothing gets on a black person's last nerve more than to hear whites argue that affirmative action has resulted in the lowering of standards and the hiring of unqualified workers. Long before a black person ever stepped foot inside a white company as anything other than janitor, charwoman, or elevator operator, whites were there screwing things up.

Whites were making all the business decisions. The industrial revolution was fought and won by white Americans. Blacks have been told that whites made America rich. That all that is great about this nation is *their* brain trust. I'm willing to give credit where it is due. But if that were the case, are we then to believe that all that is wrong with the corporate world today is due to the arrival of a few million blacks?

At a diversity meeting at my company a few years back, I took issue with a white colleague who feared the *Times* was lowering standards in the interest of diversity. I told my colleague that he owed the publisher an apology.

"Do you honestly think that Arthur Sulzberger would go out and intentionally hire unqualified people to work at this great paper, and not only hire them but continue to employ and pay people who aren't capable of doing the work? Where does it say in any law that an employer has to hire unqualified people and pay those unqualified to do the job?"

I was on a roll.

"It is an insult, sir," I went on, "to suggest that the publisher of this newspaper or, for that matter, the presidents of Harvard and Yale Universities would be so petty as to jeopardize the future of these distinguished institutions, some of which have survived for centuries, because they want to be politically correct."

My colleague said I had misunderstood his comment.

No, he'd made himself perfectly clear.

Whites want us to think that blacks, who comprise no more than 8 percent of the working population, have ruined these great institutions. I say this: If black folks are that powerful that our mere presence is ruining centuries-old institutions, then there are a few other institutions, starting with Congress, that we'd like to ruin! There were billion-dollar deficits, white-collar crime, Chapter 11 proceedings, and incompetence in the corporate world and beyond long before blacks arrived on the scene. So now we're to think that all that ails the country is our fault.

I figure this way: Anyone who would allow a small minority to come in and lower their standards deserves what they get.

No. Affirmative action gives whites an excuse for failure. It's not easy telling your spouse or a relative or the neighbors that you didn't get that promotion you wanted because the boss thought you were a screwup who couldn't do the job. Much better to say you didn't get the job because they decided to promote a black woman over you for affirmative-action reasons.

Think about it. Nobody wants to say they didn't get into Harvard because they were too dumb. They'd rather say that it's because all those dumb black kids were being admitted.

"You would think blacks were just taking over these places," said Curtis Barnes, my brother-in-law. "I'm surprised at how selfish whites can be. They are the overwhelming majority in the workplace, especially in positions of authority. What are they afraid of? If every black person in America got a job through affirmative action, that would only be about 10 percent of the population. That's a small piece of the pie, and they don't want us to have that much."

Now that I've gotten that off my chest, I have to confess: I'm ambivalent about affirmative action. We are not—and never will be—a color-blind society. In order to make us behave like we are, Big Brother has to step in and threaten punishment. But right now, affirmative action is doing more harm than good.

In simple terms, an affirmative action is a positive step.

If your parents and grandparents are graduates of Harvard University and major contributors to the university's endowment, yes, that should be considered when you apply for admission there.

If you got your job because your father knew the boss and put in a good word for you, fine by me. If you don't know how to do that job, I've got a problem with that.

I have no problem with the growing number of white coaches in the National Football League and the National Basketball Association because blacks represent the majority of those playing in the two professional sports. It seems only fair that if 90 percent of the players in the NBA are black and 70 percent of the players in the NFL are black, then a positive step would be to diversify the league by bringing whites into the front office. My problem with what's going on is that some of the whites being hired aren't as qualified as the blacks being overlooked.

I understand the frustration a white firefighter or police officer feels when he studies hard to pass the sergeant's examination and exalts in receiving a high score that places him at the top of the list for promotion, only to see a black who scored much lower than he did get the promotion.

"Those are the cases that bother whites the most," said Mrs. Lewis of Queens. "It's those cases where you have a white student whose grades and test scores, college essay, and extracurricular activities are better than those of a black student, but the black student gets in and the white doesn't because the university has some quota to meet."

My main problem with affirmative action is that it has come to mean that the only reason you got the job or were admitted to this or that school was because you were black and not because you were qualified or good enough to get in on your own merits. Most blacks will tell you that they may have been hired or admitted because they were black, but they succeeded because they were good. Are we to believe that before affirmative action, every white student admitted to Ivy League schools got there based solely on merit? Yeah, I'd like to buy a bridge!

Mediocrity isn't race specific. It's not always the cream that rises to the top. Maybe it's just Cremora. Looks like cream, tastes like cream. It ain't cream.

The problem with affirmative action, says my friend Michael Anderson, is that it is a racist system. "America is so racist that the only way it could solve racism is by more racism," Michael told me. "The only way to change it is to decide which things are unfair and do away with them."

The state of New Jersey wrestled with that exact issue in 1995. The case involved a white teacher and a black teacher who were hired by the Piscataway school district on the same day and had similar qualifications. When one had to be laid off for budgetary reasons, the school board dismissed the white teacher, claiming it was acting in the interest of racial diversity. Thanks for nothing.

Of course, the white teacher sued. The black teacher would have sued if she were the one fired. The school board took the low road and had the audacity to claim it had taken the high road in the interest of affirmative action. New Jersey's black residents were as appalled by the decision as whites. Civil rights groups tried to convince the school board to settle the case, but no, the board pursued the case all the way to the Supreme Court before agreeing to settle out of court for $433,500. Maybe the school board figured the money paid to settle the case had less impact on the school budget than what the board might have saved by firing the teacher. You do the math.

Bakke was another matter. Where is he now? Who knows? And I certainly don't care, although the last I knew of the infamous Allan Bakke, he was working as an anesthesiologist in Rochester, Minnesota.

Bakke, who is white, claimed he was discriminated against in 1973 when his application for the University of California Medical School was passed over in favor of minority applicants. Bakke filed suit and his case reached the Supreme Court. In 1978 the Court gave Bakke a partial victory and struck down the special admissions program at the school but also held that race could constitutionally be taken into account based on adequate findings of past discrimination.

The Court's ruling was supposed to clear up the affirmative-action debate. It only added to it.

Finding a better way, something other than affirmative action, won't be easy. We can't go back to the days of Who do you know? and How do you know them? Nor should we just throw up our hands. That's what the American Society of Newspaper Editors appeared to do in 1998 when it pushed back its original timetable to reach ethnic parity in the newsroom. In 1978 the society, made up of 860 editors from around the country, aspired to have the makeup of its news room reflect the ethnic diversity of America by 2000. But having combed the countryside for nearly two decades in search of "qualified" minority applicants and finding fewer than they envisioned, the society agreed to push back its timetable to 2025. Maybe two more decades would do the trick.

I shouldn't be so cynical. ASNE is one of a few professional organizations to agree to goals and timetables to bring its workforce in line with the changing face of America. My problem is the group may have been too quick to throw in the towel. Maybe we all should realize that we've only been at this for thirty-six years. Let's not go so quickly into the night.

Raise the roof and the glass ceiling. Redefine success. If we do nothing more, stop blaming the victims: indeed, affirmative action has made victims of us all.

Is He Black? Is She Black?

Whites are not supposed to be sensitive when it comes to racial matters. They aren't supposed to "get it" when it comes to the "black thing." Often when blacks are told about someone in a position of authority who has over a sustained period demonstrated a sensibility to our people, we assume the person couldn't be white.

Oh noooo! They actually are out there recruiting blacks to work at their company? Must be black. They're the vice president of the company and invited one of us to their home for dinner? Naah. They couldn't be white.

When we're uncertain about someone's race, we pick up the phone and call a brother or a sister who might be in the know. "Hey, do you know if so-and-so over at such-and-such company is white? Sounds like she could be black, but you never know."

Whites, either through deduction or experience, know blacks do this and it bothers them.

"What difference does it make whether I'm black or white," said a white man who was recruiting for a mass-media company at a job fair held at Howard University in Washington. He was sitting in a booth opposite me during the fall of 1998. Both he and I had been sent to the prestigious black college to look for promising young college students to work for our companies. During one slow moment, we struck up a conversation. He said he noticed that the black students were far more willing to approach me than him. From there we went into the reasons why, and the conversation eventually led to the common practice among blacks of trying to figure out the race of the person they may have to face.

"Perhaps it's because we want to prepare ourselves for what may be," I tried to explain. He, however, saw it differently.

"I get the sense," he said, "that blacks automatically assume if the person they have to meet or interview with is white, they're going to be given a hard time. I mean, I feel as though blacks assume that because we're white, we're not going to be nice to them. I think I'm a nice person, period. It's not like all blacks are nice and all whites are mean."

True. But tell that to some black people I know.

Whites Have It Made

My father had this saying: "It's a white man's world; I'm just passing through."

There are still blacks who feel this way—that whites have it made and have nothing to worry about. With all the opportunities they are afforded, they should excel. After all, they're white. If they do fail to achieve, then it's just "whiteness gone to waste."

As such, there are blacks who believe that blacks should concentrate on helping each other. Whites will take care of their own. Why should we be mentors to them? They've got more than their share. Why should we blacks give them a leg up? Before we know it, they'll be at the top of the ladder looking back down on us.

But isn't that another form of discrimination?

When I spent a month at Duke University, I was approached by a group of white students who asked if I would have lunch with them.

"We heard you wanted to talk to some of the students here, and we wanted to ask you some questions," a white female graduate student asked me. She said there were four or five students, all white, who were interested in picking my brain. "But we weren't sure if you were just talking to the black students."

Her honesty caught me off guard. Initially I was taken aback that she and the other white students would think such a thing. I wasn't like that, I assured her.

"Half of my mentors have been white," I said. "It wouldn't be right for me to turn around and discriminate as to who I will and will not mentor based on the color of their skin. I would dishonor my own white mentors by refusing your request. Besides, I wasn't raised that way."

So we had lunch. Four white female students and one white male. They wanted to know about the professional world, about New York, about working at the *Times*. About whether you could have a career and a family.

I asked them about their backgrounds, where they were from, what they hoped to do in life, what campus life was like.

They told me of their ambitions, and they also told me about their fears.

"We're supposed to succeed, you know," said the graduate student, a tall, attractive sandy-haired blond, with blue eyes, a southern drawl, and ambitions of working as a television correspondent. "My father's a physician. My mother's a lawyer. I went to private schools all my life, and I'm at Duke. And I'm a white female. I'm not supposed to fail."

The male student, a senior from Northern California, said he didn't go to any parties on campus, had only been to one basketball game in the four years he was at Duke, and didn't date, because he didn't have time.

"I'm trying to graduate with honors," he confided. "My parents honestly believe they gave birth to a doctor, and I'm supposed to fulfill their dream. I don't want to disappoint them."

He felt his female coeds had a better chance of getting breaks in life because of their gender. They, in turn, felt that as a white male with a Duke degree, especially one with honors, he didn't have anything to worry about. (His world; the rest of us were just passing through.)

Talk about opening my eyes. Here before me sat children of white privilege. And yet, at that moment, I did not envy them. They were right. Because they were white they were expected to achieve. Black folks are expected to fail. If we achieve, it's viewed as a quirk of nature.

I was one of the children of Dr. King's dream. One of the beneficiaries of the civil rights movement, I came along at the right place and time in history. My family, my teachers, my neighbors, and members of my church convinced me that I could go far, if I stayed on the straight and narrow path. I didn't want to disappoint them. So when the guy I was dating in high school wanted more than a kiss, I broke off the relationship. I seldom dated during my four years at Howard, steered clear of the fast crowd, the drug crowd, and slackers. I worked on the student newspaper and the yearbook, worked part-time at Howard's radio station, worked as a student teacher. I was considered a good girl.

But I managed to have some fun. I participated in my share of anti-Vietnam demonstrations, embraced the black power movement, and attended my fair share of parties. I wanted to be a journalist, but if I didn't make it, I wouldn't have felt as though I had failed; my family wouldn't let me. They were proud that I graduated with honors and had been accepted at Columbia University's graduate school. In their eyes, I was already a success.

Success, however, is relative. Whites think that a black person making anything above $25,000 a year is a success. Whites don't consider themselves successful unless, for starters, they're doing better than the majority of blacks.

The white students who spoke with me that day understood that they were expected to do better than their parents and better than blacks like me. My advice to them was to enjoy what might very well be the best years of their lives.

I also felt a bit sad for them.

The Duke students did not fit the profile of the troubled youth in America. The media has focused so much attention on the plight of troubled minority youths, that we fail to think that other young people might also want our help. What could possibly be bothering privileged little white students at Duke? What worries could they possibly have?

Try living up to white America's expectations.

Right Don't Wrong Nobody

My mother used to say that "right don't wrong nobody. When you're wrong, you're wrong!" She raised us with a sense of righteousness and justice for all. We were taught that when someone did something wrong, we shouldn't let them get away with it simply because they are black.

Many whites cannot understand why blacks refuse to condemn members of their own race who do things that are obviously wrong or offensive. Our failure to do so has often caused strain and tension among whites in the workplace, school, and beyond.

Nothing pointed out the schism more so than the Clarence Thomas hearings. White women could not understand why black women were so openly critical of Anita Hill. The white media waited for the so-called black leadership and prominent blacks to rally to their editorial cries that Mr. Thomas not be confirmed. Although several blacks fought against the confirmation of Justice Thomas, the popular consensus among blacks was that whites were using a black woman to keep a brother down.

Besides, blacks noted, Mr. Thomas stood accused of acts no worse than some committed by members of the all-white male Judiciary Committee that was standing in judgment of him.

The Thomas case should have prepared white America for what came next: Marion Barry!

Barry, the mayor of Washington, D.C., had been caught, on videotape in a sting operation, preparing to smoke crack with a black female friend.

While the nation responded with shock and horror, many black Washingtonians accused the government of a setup. Again, blacks saw whites using a black woman to get another black male. Barry was seen as a pariah to whites. But blacks turned him into a kind of folk hero.

When he was heard on the videotape saying, "The bitch set me up," his legendary status grew exponentially. T-shirts with his remark sold out on arrival at some street vendors.

Barry was later convicted and served time in prison, but blacks in Washington got a final say on the matter when they reelected Barry to office four years after the incident and two years after his release from prison.

"Now, explain that to me," a white friend said, asking me about the Barry case. "Why would blacks in Washington put a man like that back in office?"

I didn't agree with what was done, but I understood the logic behind it.

Understand. Someone caught on tape preparing to smoke crack isn't my idea of a leader. But, as many blacks pointed out in interviews for the book, the president of the United States has been investigated for everything from fraud to adultery to lying under oath. Barry's misdeeds pale in comparison, they told me.

And another thing. Washington has always been a predominantly black city. Approximately 63 percent of the population of the nation's capital is black. The District of Columbia is not considered a state but federal property. Residents pay taxes, but do not have a voting member of Congress. For years, the people of D.C. have sought home rule. Congress has refused. We—and I speak now as a native—did not have a mayor until the mid-1960s. Walter Washington was chosen to serve as a ceremonial mayor of Washington. We had our first election in 1974–75 to choose a mayor and city council.

Blacks in Washington are convinced that if the population majority were white, home rule would have been granted a long time ago.

They elected and reelected Marion Barry because they not only wanted to have a black in the District Building, which serves as city hall, but also knew his return to office would be an affront to white Washingtonians.

"Every time a well-known black person does something that whites don't like, they want black leaders to disavow that person," Mr. Frazier, a black Washington executive, said at one focus group in Washington. "I think black leaders have grown weary of that. No other race is called upon to address the actions of one of their own kind who misspeaks or engages in perceived discriminatory or racially insensitive behavior. So why should blacks be held accountable for the actions of one?"

Just Tell Me What to Call You

Many whites expressed confusion over the issue of racial identity.

"I can't figure out what blacks want to be called," Calvin D., a white man in his late fifties, said over a cappuccino in a coffee shop on Sunset Boulevard one day. "By the time I was old enough to remember, whites were no longer referring to blacks as colored. I personally stopped using Negro in the 1960s and started referring to blacks as blacks. Then came the African American reference. But some say they prefer to be called black Americans. It's confusing to us. I think most whites will call blacks whatever they want to be called, but we're not sure because it appears that many blacks aren't settled on this issue."

He's right. And it's one of those little things that can send white folks up a wall. 'Cause we tend to blame them if they take liberties.

This much can be said: *Colored* and *colored people* are definitely no-no's. Elderly blacks, those raised at a time when the term *Negro* was commonly used as our racial identity, still prefer its use. They see in it a historical legacy that transcends time, space, and happenstance.

Blacks born after World War II are more likely to prefer the term *black,* which came into popular use during the militant 1960s as a defiant response to *Negro,* a term bestowed upon blacks by whites. *Black* is a term we chose for ourselves, and most self-respecting former Negroes readily embraced its usage.

African American began to surface in the 1980s. It coincided with an ethnic boom in America. A new wave of immigrants began to challenge America's melting pot by adding their own ethnic ingredients and flavors. Much like the Italians, Irish, Poles, and Jews had done before them, blacks who had never shared the same sense of national identity as whites began to claim their rightful heritage by adopting *African American.*

There are blacks, however, who have problems with that.

"Slaves were brought to America from the Caribbean and South America," said Alice Beckwith, my brother-in-law's sister. "There were also freed blacks who came to America from Europe and those who came as indentured servants. Not every black person in America came here from Africa. We may be Euro American, Carib American, African American, South American."

Some blacks resent the term *African American* simply because they believe whites have given it their seal of approval based on the opinions of a select group of black leaders and black intellectuals.

Just because Jesse Jackson and Al Sharpton say blacks want to be called this or that, whites assume they are speaking on behalf of all blacks and, therefore, it is so.

The media picks up on it, and before you know it, every black person in America is being referred to as African American.

8. It's Not Just a Black/White Thing

After my story appeared in the *Times,* I was approached by some of my Latin and Asian colleagues who wanted me to know that many of the daily injustices suffered by blacks are also suffered by other ethnic groups.

"It's not just a black/white thing," said William Celis III, a friend and former *Times* colleague. Bill is Mexican American, but you can't necessarily tell that by looking at him. He's six feet, medium build, with blue eyes, dark brown hair, and a complexion that becomes flushed during embarrassing moments.

"I've had people tell me, 'You don't look Hispanic.' Or they tell me that I don't act like most Hispanics. And what kind of last name is Celis, anyway," he said. "Now what does that mean? I guess because I'm not short, eating beans and rice, and dancing the Latin, I don't fit the stereotype."

Repeatedly I was urged by my ethnic brothers and sisters to include a chapter in the book that dealt with this issue.

Americans of different ethnicities took offense when President Clinton failed to appoint members of other ethnic groups to his task force on race. They also demanded that the task force's focus be expanded to look at discrimination based on ethnicity. Mr. Clinton defended his decision by saying that America's troubled past was largely a black/white issue.

"If we can address the problems between black and white Americans, then we will be better equipped to deal with discrimination in other areas," the president was reported as saying.

But that was little consolation to members of those groups, even though the president did appoint a Native American and a Hispanic to the panel.

In interviews with Asians and Hispanics, I was told of daily interactions between them and white Americans that left these minority members smarting.

One Asian female reporter told of attending a press conference in which she was told by a white official: "We don't talk to the Chinese press."

"I'm not with the Chinese press," she told him. "I'm with the *New York Times.*" Upon hearing that, his attitude and posture toward her changed immediately. There wasn't anything she could ask for that was not obliged. "But why did I have to invoke the *Times* name to get that kind of respect? And why did he automatically assume that because I was Chinese, I must have been with the Chinese press?"

David Gonzalez, a Puerto Rican–American who is a foreign correspondent for the *New York Times,* told jokingly of how often telemarketers call his home and assume he can't speak the language simply because he has a Hispanic surname.

"They speak English very slowly, as if I won't understand them," he explained.

Now, David, being the New Yorker that he is, speaks fast, so he doesn't have time to listen to some slow-talking telemarketer trying to sell him something over the phone. When he responds in his most astute Yale-educated diction, the caller seems surprised.

A Hispanic female friend recalled an incident that occurred when she worked at a newspaper in Texas. For more than a year, a white male editor who sat directly across from her always passed by her desk without so much as a "Hello," "Good-bye," or "How are you?" When word spread throughout the newsroom that she was leaving the paper for a job as an editor at the *New York Times,* she said the same white male editor who had ignored her all year approached her at her desk and wanted to know "my life history."

"Everything from where I was from to what high school and college I attended," she recalled. "I knew he was trying to figure out how I ranked to get a job at the *Times.* We talked for a good ten or fifteen minutes. When we finished he said something about how sorry he was that we didn't get to know each other before. *Huh.* And who's fault was that?"

I spoke with members of ethnic groups who said they also sense that whites do not understand why they sit together or talk among themselves. What these minorities see as basic human nature, whites find offensive, they told me.

"Especially if we begin talking in Spanish," said Egardo Orellana, who was born in El Salvador but moved with his family several years ago to the Washington area. He is a maintenance worker at Sibley Hospital outside Washington. "They always assume we're either talking about them or trying to exclude them from our conversation. But *they* can sit at a table and laugh and joke, and that's OK. They also get upset when the blacks and Hispanics seem to get along. I guess they want to see us at each other's throats. But when we're united and stand together, the whites feel threatened by it."

In focus groups in Los Angeles, New York, and Washington, several Asians and Hispanics spoke bitterly about the numerous times they had been subjected to humiliation and ridicule based on white stereotypes.

"You know, we're all supposed to be brilliant; you know that, don't you," a thirty-one-year-old Japanese American woman told me during a brief chat in Hartsdale, New York, a suburb that has a sizable Japanese population.

She laughed. Mirito is smart. She graduated from Yale University and works as an economist for a major foundation. She sought anonymity, she explained, because she is the only Japanese American at the foundation and thus could be easily identified. Assured her identity would be protected, she continued.

"I went to a predominantly white high school in New Jersey, and every time the teacher asked a question that none of the white kids could answer, everybody in the class would expect me to have the answer. I may be smart, but I'm not brilliant. I'm no Einstein by a long shot. But I was expected to supply the answer, and if I didn't, I would see this look of disbelief etched on my classmates' faces. That followed me through college and into the workplace. 'She's Japanese. You know, they're supposed to be smart.'"

They're also supposed to be self-deprecating and void of emotion.

"Oh, yeah, I get that, too," Mirito said. "But I don't fit that stereotype. I'm outspoken and outrageous, and that comes as a shock to some of my colleagues."

Mr. Celis, whose speech is void of either a Spanish accent or Texan drawl, gets the old "You don't sound Hispanic."

"All the time," said Celis, who speaks softly. "That's because America ben bery bery good to me."

He was laughing, but he admitted he doesn't find that kind of ignorance funny.

While Japanese are expected to be smart, Hispanics say they are often viewed as dumb.

"Maybe it's because of our accent or because you don't see a lot of Hispanics in leadership roles or positions in America," said Alex Ortiz, who is Puerto Rican but was born and raised in New York. "Whatever the reason, people think Hispanics lack education. I've had whites ask whether English is my second language. No, English is my first language."

Mr. Celis has also heard the stereotype that Hispanics lack intelligence.

"Unlike black Americans, we have not been a people

who has recorded its history, so our achievements aren't as widely known as other groups', and that leads to this perception that we aren't intelligent or don't strive to greater heights. In the mass media, Hispanics are portrayed in movies or on television playing subservient roles, which has further fueled the stereotype of my people as silly and dumb."

Blacks have played to the stereotypes about ethnics.

Blacks have referred to Hispanics as "hot tempered." We've bought into the belief that Latin men are the macho, jealous, possessive types who want their women to be beholden and subservient to them.

"Oh, and you know we're supposed to love those loud colors," said Juana Bettancourt, a cousin from my mother's side, whose family emigrated from Cuba. "Reds, greens, oranges... bright, bright, bright!"

One little thing—or maybe it's a big thing—that riles most Hispanics is the assumption by many native-born black and white Americans that everyone of Spanish descent is from Puerto Rico or Mexico.

"I'm from El Salvador," said Mr. Orellana. "But the minute people hear me speak, they think I'm Puerto Rican or Mexican. I think blacks are a little more sensitive to that issue than whites, though. Blacks will ask, Where are you from? Whites tend to ask, Are you Puerto Rican? Another thing that bothers me is that regardless of where people of Spanish origins are from, most Americans expect all of us to share the same cultural tastes, like the same foods, and to all be Catholics."

He forgot to add, play baseball.

Angel Franco, a photographer at the *New York Times,* took up golf for that very reason. Over the ten years he has played the game, Mr. Franco has lowered his handicap and now consistently shoots in the nineties. Whenever he plays a round of golf with whites, especially those unfamiliar with his game, he draws stares, not only because he strikes the ball with such accuracy but because of how far he can drive the ball.

"I remember once a white guy asked me what got me into golf," said Mr. Franco. "And I told him: 'To see the look on your faces.' "

And another thing.

It's my understanding that, like black Americans, people of Spanish origin didn't just start calling themselves Hispanics. President Jimmy Carter was the first to officially use the term.

Unfortunately, it's rather broad. The dictionary says the term *Hispanic* encompasses all Spanish-speaking peoples in both hemispheres and emphasizes "the common denominator of language between communities that sometimes have little else in common."

The same way white folks tagged us with the *Negro* label, Carter placed the *Hispanic* label on people from Spanish-speaking countries in the Western Hemisphere.

Asians believe that they, too, are viewed as a monolith. The old stereotype that they all look alike is commonly applied to Asians. Blacks and whites confided to me that they cannot distinguish among Chinese, Japanese, Koreans, or Asians from the Pacific Rim. Too many of us have bought into the idea that Asians are "sneaky people."

I've heard that since I was a child.

I assumed it grew out of World War II and the bombing of Pearl Harbor by the Japanese and the estranged relationship that then ensued between America and Japan. An older generation of Americans said it had to do with the shape of Asians' eyes and their quiet ways. I used to think people were joking when they said the chicken served in Chinese restaurants is really cats until I'd heard it once too often and began noticing that some Americans shy away from ordering the chicken dishes in Column A.

Those who say it isn't just a black/white thing are right. But many blacks feel that members of other ethnic groups have often received preferential treatment at blacks' expense. Some of us blacks are disturbed that Hispanics, Asians, and members of other minority groups have benefited from civil rights laws intended to redress wrongs committed against blacks.

"We—blacks—were brought here in chains, forced into servitude and second-class citizenship, and chastised for refusing to get over the past and on with the future," said Mrs. Milton. "Hispanics and Asians do not share that history with us. You can't be a minority in America and not be subjected to some kind of prejudice or discrimination. But you can't compare the plight and history of black Americans to other ethnic groups, because what we suffered was cruel and unusual."

"You know the old saying: 'If you're white, it's all right; if you're brown, stick around; if you're black, step back.' It's *anything* but black," said my sister-in-law Joanne Williams. "I've seen immigrants come here, barely knowing the culture or the language, and move faster and further than blacks born and raised in America."

Like other blacks, Joanne says her blood boils when she encounters newly arrived immigrants "who have the audacity to look down on black people like we're dirt."

"You go into a Korean grocery store and they follow you around like you're a common thief," said Joanne. "They want you to shop in their stores, paying twice as much for products we could buy at the Safeway, but they don't want to give us due respect. That's why I don't go in their stores."

In an editorial published in *USA Today* in July 1998, Camille Cosby, educator, producer, mother, and wife of Bill Cosby, said that Mikail Markhasev, a Ukrainian émigré convicted of killing her son, Ennis, was a racist and that America taught him to hate black people.

"I believe America taught our son's killer to hate African-Americans," Mrs. Cosby wrote. "Presumably, Markhasev did not learn to hate black people in his native country, the Ukraine, where the black population was near zero."

Mrs. Cosby said her son's killer learned to hate blacks through racism and prejudice "ominipresent and eternalized in America's institutions, media and myriad entities."

If Markhasev had come from Haiti, the only way the boy would have been allowed to enter the United States would have been through the back door.

"They're importing white people," my niece Angelique likes to say. "I'm not kidding, Aunt Lena. Whites in America know their days as the majority race are over. Their population is shrinking and people of color will become the new majority. Rather than see that happen, they're bringing in people from the Soviet Union or Eastern bloc countries. These pale, blue-eyed Slavic types. It's true, I tell you."

In truth, immigration didn't seem to become a problem until it became a "black thang!"

Congress didn't really begin to clamp down on immigration until all those Africans, Haitians, and dark-skinned Cubans starting arriving, uninvited, on American soil.

Under new immigration laws, *aliens*—the term used in the law—face deportation if they so much as spit on the ground. I do not exaggerate. It's called moral turpitude, and under a stricter deportation law approved in 1996, green-card-carrying immigrants convicted of a felony or misdemeanor can be deported.

Blacks know that every other group—but us—voluntarily came to America in search of a better life. Black folks were brought to America blindfolded and in chains, against our will.

The Statue of Liberty wasn't symbolically reaching out and welcoming us with open arms. We couldn't see old Miss Liberty there from the hulls of slave ships, so that loving image isn't in our collective memoirs.

Blacks, however, are believed to have more in common with whites than members of other ethnic groups, even those born in America.

My former housekeeper, a Hispanic woman, told me she'd been told by the white woman who runs the New York cleaning agency she works for that many white families would prefer to have a black housekeeper than a Hispanic one.

"She said that's because white people think blacks know how to speak English, are legal citizens, and know the culture," said Nina. "She said the whites say they can't understand our English and we don't have the same values and customs that native-born Americans have."

What Are You?

Americans of mixed-race heritage complained to me that their lives have been defined, all too often, by what they are not.

"What are you?" Susan Stovall has been asked most of her thirty-something years on this earth.

Ms. Stovall is the daughter of Charlayne Hunter-Gault, who is black and shares a place in history as the first black to desegregate the University of Georgia, and Walter Stovall, a white southerner who met Charlayne while covering the story for the Associated Press.

Raised in New York and now living in Los Angeles, Susan brought together a group of friends in her apartment—two of whom, like Susan, are biracial, and a third who is raising a biracial child—for one of my focus groups.

"It was right after the Howard Beach thing," said Susan, referring to an incident in which a young black man was struck and killed by a car while being chased by a group of white youths from Howard Beach, a largely Italian neighborhood in Brooklyn. "I'm standing in line at a movie theater in New York, listening to my Walkman, and I accidentally bumped the arm of this black man, who was with another black man and two black women. I didn't realize I'd bumped him and I'm listening to my music, oblivious to the world, when I hear one of the women say, 'Damn, don't nobody have any manners anymore.' Then I hear: 'This ain't Howard Beach, you know.' The two women are looking at me, snickering, and I felt like I was back in third grade, at Public School 165."

Back then, Susan recalls, she was chased home from

school by groups of black or white or Hispanic students "because I didn't belong to any one specific group."

"I wasn't black, I wasn't white, I wasn't Hispanic," she explained. " I was forever being asked, 'What are you?' 'You don't look black.' 'What's your nationality?' "

That day, standing there in the movie line, it all came back to Susan: the teasing, the probing questions, the ridicule. She said she apologized to the man whose arm she had bumped, and then told the group that she appreciated their anger about Howard Beach, because she, too, was angry.

"Then I said that if they had taken time to look at me—at my lips and my behind—maybe they wouldn't have been so quick to judge."

Roslyn Myles, whose ex-husband is white, says she worries that her mixed-race son may soon come face-to-face with the ugliness of racism. "I know he will hit this wall and have this ugly day," she said. "I watched him the other day, playing with his father. It was no big deal. Daddy's white and Mommy's black. No big deal. But one day he will go to school or be at the playground or at the mall, and that is going to be a big deal."

Michael Jackson declares in a song: "It doesn't matter, if you're black or white." We wish.

A telling vignette of that was revealed to me several years ago, when my then eight-year-old niece invited a group of her parochial-school friends over to our home in Washington for a birthday party. It was a mixed group of boys and girls. For some reason the kids started talking about who they were in terms of race.

"I'm black," one little girl in cornrows declared.

"I'm Dominican," a brown-skinned little boy with dark curly hair volunteered.

"I'm white," said another little boy, who definitely looked the part.

Finally the group came to a little girl with thick shoulder-length pigtails and a complexion the color of coffee with a tad too much cream.

"I'm, I'm . . . mixed-up," she said, squirming in her chair.

My sister and I both had a mixed-up look on our faces. Unable to contain her curiosity, Ada asked what the little girl meant.

"You know, mixed-up, black and white," she said, looking at us through big brown eyes that seemed to say: *Shouldn't that be obvious to you two?*

"Oh," we said, and slithered out of the room.

Looking back now, it was a rather interesting way of putting it. There are blacks who might say my niece's little schoolmate was right, literally and figuratively. Many blacks believe that children of interracial couples often go through life . . . mixed-up: unsure of who they are or living a life of neither here nor there, never feeling a sense of belonging in either race.

Whites are often surprised to learn that many black parents say they are more inclined to raise their mixed-race children as black rather than as white.

"It's easier on the child, I think," said Adrian Miller, whose mother is black and father, Puerto Rican. Adrian is a light-brown-skinned young man who has worked in the entertainment industry in Los Angeles. "If the child doesn't look Anglo-Saxon but is raised as white, then they'll go through life always being challenged. Maybe it goes back to the one-drop rule. Unless you look white, whites aren't necessarily going to accept you. Blacks, however, will accept people who don't necessarily look black."

When they do that, however, a child is forced to reject his or her white parent. Every time we demand that you declare, one way or the other, what you are, we deny that which we are not.

Besides, not all who look half-and-half really are.

Shawn Kennedy, one of my closest friends, is black, but you couldn't tell that by looking at her or her identical twin sister, Royal.

A reporter at the *Times,* Shawn has felt the sting of racist comments made by whites to her about blacks and has watched their reactions when she tells them that she is black. Some think she's playing a cruel joke. Others don't know what to think and simply walk away.

Shawn has also been slighted by blacks who assume she is white and do not know that she is married to a black man because she is a black woman. There have been moments in her life—I'm sure more than she cares to remember—when she has had to explain herself. Anytime it appears that she is getting preferential treatment as a minority, she finds herself explaining why. She's been asked, "What are you?" more than she cares to remember. Yet she still responds politely.

In the more than twenty years I've known her, Shawn has refused to exploit the uncomfortable positions she sometimes has been placed in because she does not fit the image of an "authentic Negro." White men who wanted to date her were told up-front that she was black. Black men who wanted to date her, thinking she was white, were set straight.

It really shouldn't matter what she is. But Americans like to know where they stand when it comes to race. We want to know who and what we're dealing with. We may be racists; we just try not to show it in public.

Conclusion

Morris Dees. John Brown. Edward M. Kennedy. Paul Newman. Andrew Hacker. Joe Feagin. Branch Rickey. Bill Clinton. Claudia Payne. Judy Twersky. The Witches.

What do these white Americans all have in common?

They are "white folks with some currency."

The saying—which dates back to the 1950s—was used to describe whites who have demonstrated a racial sensibility, either through actions or deeds. Acquiring currency among blacks isn't an easy task. Many whites have made the mistake of assuming they have currency in the black community simply because they have black friends.

Keri Eisenbeis, a young white woman and former girlfriend of Chuma Hunter-Gault, the son of Ronald and Charlayne Hunter-Gault, heard me use the phrase "white folks with some currency" and said she wanted to know how she might acquire some.

She already had.

Heads turn and eyes roll when they see Ms. Eisenbach, a tall attractive brunet, walking arm in arm with Mr. Gault, an actor with brown dreadlocks and blue-gray eyes. She, however, seems unfazed by the whispers or stares and appears as comfortable in the company of black strangers as she does with whites she's known for years.

"When I was in South Africa, Charlayne and Ron would be entertaining black South Africans at their home," Keri said, wanting to know whether the incident meant she had currency. "Whenever the conversation turned to racial matters, Charlayne and Ron would be going on and on about white racism and white this and white that, and the black South Africans would be staring at me, trying to gauge my reaction. One time one of them took Charlayne aside and said, 'You must not say these things in front of a white person; what must she think.' And Charlayne said, 'Are you talking about Keri? Oh, Keri doesn't count.' "

It's another one of those little things that can be confusing to whites. How blacks can forgive some whites for their mistakes, yet the same mistake by others is viewed as a cardinal sin.

Ted Kennedy curries favor among blacks because he is considered a champion of civil rights. Vice President Al Gore inherited currency from his father, the late Albert Gore Sr., who was one of the few southern congressmen to openly support civil rights issues. Gore kept some currency by showing support for black causes.

I have always loved Paul Newman. My feelings toward the actor partially stem from his work as a screen actor, but also because I remember seeing him at the historic March on Washington in August 1963. I was thirteen years old at the time. Newman was among several white celebrities—Marlon Brando and Peter, Paul, and Mary among them—to participate in the march. In doing so, they publicly aligned themselves with the fight for civil rights.

Eleanor Farrar, a white woman who shares the summer home in Betterton with Ms. Camper and Ms. Schneider—both black—has worked at the Joint Center for Political and Economic Studies in Washington and has lived, socialized, and agonized with blacks of various ages, classes, and regions. During the focus group in Betterton, Ms. Farrar took me aside and told me that "getting over past racial injustices was easier said than done."

She said, "Even when you think you've gotten over it, something happens and you realize you haven't."

Just how far Ms. Farrar's currency goes was illustrated during my weekend stay at their home. At one point Ms. Camper, referring to the blacks present, said, "All of us here think such and such." Ms. Farrar never said a word, even though the other blacks in the room were convulsed with laughter. Diane finally realized her mistake.

"Oh, we consider Eleanor one of us," Ms. Camper said with a wink.

I count among my closest friends, Jill, Claudia, Judy, and a group of female colleagues at the *Times* known as the Witches: Nadine Brozan, Georgia Dullea, Bernadine Morris, Kathleen Teltsch, Anne-Marie Schiro, Andrea Skinner, and Shawn Kennedy. All but Andrea, Shawn, and me are white. (By the way, if you're wondering why we call ourselves witches, the story goes that one of the editors once told another editor that she wasn't to tell those . . . anything. When the editor told those . . . what was said, she told them that the word rhymed with *witches*.)

Over the years these white women—there are white men in the group, but we women have a special bond—have stuck by me through many trials and tribulations. They have offered everything from encouragement to financial help. We have partied together, vacationed together, struggled together, cried and laughed together. They know my family, and I theirs.

Oh, we've had our differences of opinions and tastes. But they have demonstrated a sense of understanding where it counts the most: by refusing to engage in racist activity or behavior, whether in the workplace or home or social setting. When whites have done so in their presence, these women have taken a stand.

Jill Gerston and I go way back.

Jill is a former reporter for the *Times* and the *Philadelphia Inquirer,* and I have known her for nearly twenty-six years. When we first met, I thought we had little in common, except that she was a nice Jewish girl from Brooklyn and I was a nice black girl from Washington. Nice girls. That we had in common. She'd graduated magna cum laude from Vassar College and seemed destined for success at the *Times*. I graduated cum laude from Howard University and wasn't sure where I was headed but felt it wasn't in the same direction as Jill.

Still, she couldn't have been nicer. She always spoke,

offered advice—which I interpreted as white folks' tendency to think they know what's best for black folks—and offered encouragement. We became friends in spite of me.

Over the years, we have talked about race matters. We have agreed on certain things. Disagreed on others. She believes that Jews and blacks share a common bond and suffering. I don't always agree that the races share parallel lives. I tell her that she can walk down the street with her husband, and people see a white couple. But if I walk down the street with her husband, they see a black woman and a white man and they may react to it.

My friendship with Jill transcends race. So much so that when my mother finally met Jill, years after our friendship began, she told me that Jill didn't look at all like she pictured her.

And how had she pictured her, I asked my mother.

"Well, frankly, black," my mother said with a smile.

Because I was so close to Jill, my mother assumed she was black. In all the years I talked about her, it came to me that never once had I mentioned her race.

Why? It wasn't relevant. It wasn't important. Jill was Jill. She happens to be white.

I, in turn, do not have to prove my racial loyalties to Jill or her husband, Jay Newman. I, too, have acquired currency. I can talk openly with both of them about intragroup issues that were once taboo to mention to anyone outside the race. They have given me an insight into Jewish life that has opened my eyes, mind, and heart. The Christian and the Jew have prayed together and stayed together.

We all can't be friends. But we can cut each other some slack.

Maybe what the races need more than anything else is to lighten up.

Maybe we need to understand the historical baggage carried by both blacks and whites. What I've tried to do in this book is to offer little brain cramps: a tiny little tinge that, perhaps, makes you think before you act, to think less about what is intended, but more about what may be felt.

Maybe the next time a group of whites walks down a crowded street and a black man approaches, they'll try to understand that what may be going through his head may not simply be who yields the right-of-way. Maybe there's some emotional racial baggage he's been carrying around so long, he's forgotten it's there.

Maybe the next time a white woman slings her hair in my face or my brother-in-law's, we'll think she's nothing more than a flirt.

Some of us may never be a victim of redlining or racial profiling. Some of us may get our jobs or promotions based solely on merit. But if we live long enough, sooner or later, all of us, blacks and whites alike, will get nicked by one of Dr. Poussaint's micro-aggressions. One little nick won't kill us. A thousand, however...

It was, you recall, the proverbial straw that broke the camel's back.

Was it simply that people were so caught up in their own baggage that they failed to see they had placed too great a burden upon the camel? Or was it the camel's fault, as my friend and colleague Michael Anderson suggested?

"Maybe the camel should have dumped those straws a long time ago," said Mr. Anderson, who is black. "The problem we have as a people is that we are trapped in history. History can't be changed. We have to understand that and move on.

"There's no way to understand what's going on in people's minds," Michael observed. "Maybe whites have more to learn, but blacks have a pedestal that they need to get off, too."

Blacks may hate to admit it, but the brother has a point.

We expect whites to know better, because we think we know all there is to know about them—we had to. They didn't have to do the same. It's that simple.

I don't hate white people. I hate, hate.

"People have to know that racism isn't cool," Keisha of Brooklyn tells students she counsels through the Anti-Defamation League's peer-counseling program. "It doesn't look good; it doesn't sound good; don't do it."

The blacks and whites who participated in my forums said there should be more opportunities for blacks and whites to sit down and talk to each other, rather than *at* each other.

Although there were moments of tension when we first sat down, most people left—some staying long after the discussions ended—carrying on personal conversations. There were hugs, handshakes, and in some instances, exchanges of telephone numbers and business cards. I even heard a couple of times: "I'm sorry, if I..."

"You know, sometimes we do blame all white people for what a few do," said Joanne Davis, an evangelist from Washington. "I've met some really nice white people and some who don't care at all for blacks. But I don't let those things bother me anymore. You can't change everybody.

Besides, no matter how hard you try, there will always be prejudiced people in the world. All we can do is try to educate people, to show them how harmful discrimination and prejudices can be for everybody involved."

Jews have a saying: Forgive and remember. Those who believe we can forgive and forget are fooling themselves. The Jews have it right. If we heed their advice, we will forgive each others' trespasses and, in remembering them, not be condemned to repeat them.

Or maybe we should do as Lorraine Hansberry suggests in *The Sign in Sidney Brustein's Window:* "Yes... weep now, darling, weep. Let us both weep. That is the first thing: to let ourselves feel again.... Then, tomorrow, we shall make something strong of this sorrow."

I will continue to be pained by the little things. Every time a taxi passes me on the street to pick up a white fare, every time I'm seated at a table near the kitchen at an exclusive restaurant, every time I get "the look" when I say I'm Lena Williams of the *New York Times*—every time, I suffer a tinge of indignation. In time it passes. It must. I cannot afford to let it fester like a sore. I have work to do, people to meet, places to go, a life to get on with.

Afterword

"And Another Thing"

RINGLING BROTHERS sent me a clown.

The emotional scar left nearly forty-five years ago by a circus clown had long since healed. Or so I thought, until a Ringling Brothers clown with two colorful sprays of balloons arrived at my book party in Washington, D.C., in the Fall of 2000. Of the many things I wrote about in the book, the opening anecdote about my childhood experience at the circus touched people in ways I had not imagined. Judy Twersky, one of my white friends with currency, took it personally. She once worked as a public relations consultant for Kenneth Feld, owner of the Ringling Brothers Circus. After reading the book, she told the people at Mr. Feld's office about my story, and they sent a clown who handed out balloons to me and my guests, blacks and whites, young and old, no exceptions. I was reduced to tears.

In one magnanimous gesture Mr. Feld removed the scar tissue long buried in my soul. He took the time to right a wrong committed by someone else in another time and place. Longevity indeed has its place. Who would have thought that the little black girl, slighted by the circus clown in 1955, would one day grow up to know someone who knew the circus owner. I'm scared of me.

Alas, I must confess that not all reactions to my book were as positive, but that was to be expected. My intent was to get people to react, not to respond with indifference. Blacks, for the most part, could relate to the many anecdotes and experiences I shared. Many whites, on the other hand, appeared to take my observations and those of other blacks interviewed, personally. Some of the strongest criticism came from my peers in the media.

"Jews complain about discrimination, but if they were to say the things you said, they'd be called racist," a white reporter from Baltimore said during an interview.

I asked if she were Jewish. She was not. Not that I would have known. And therein lies the difference. Religion is covert. Race is overt. Most often you can tell a black person when you see one. The same cannot be said of most Jewish people.

A critic for the *New York Times* reviewed this book, writing that I felt white people smelled funny, couldn't dance, weren't hip, and didn't raise their children properly. She even had white men cowering in corners because of black men. Having just read this book about age-old stereotypes, misperceptions and misconceptions, and the scores of insensitive things we do and say across racial lines, and *this* was the message she gleaned from it. Just goes to show: you can't believe everything you read in the newspaper.

Incensed by the inaccurate portrayal of this book in the *New York Times* review, a group of black journalists began a letter-writing campaign to the newspaper and other major news organizations in order to "set the record straight."

When Elizabeth Vargas, a reporter for ABC's *20/20*, asked me to respond to the *Times* review for a segment the magazine show was preparing on the book, my answer seemed to rise from within my soul and from my ancestral past. "This is my life. I've lived it for fifty years. I've written stories about it. I spent months traveling the country interviewing dozens of black Americans about the daily injustices they suffer across racial lines. And what? One white woman writes that it isn't so, and that alone negates my life! It negates the experience of the blacks I spoke with. You want to know one of black peoples' worst fears?" I asked Ms. Vargas. "It's that the word of a single white person carries more weight than a chorus of black voices. It's been that way for centuries and apparently nothing has changed."

She smiled. Point made. Perhaps, point taken.

As I traveled the country promoting the book, I was approached by blacks and whites alike who wanted to share personal stories of racial affronts. A white woman in

Atlanta, who worked at a social services agency, said a black co-worker accused her of racism because the white woman is pro-choice.

"Can you figure that one out for me?" she asked me during a reading at a local bookstore. "I support a woman's right to choose and that makes me a racist?"

I, too, was perplexed. I told her that I'm pro-choice, but not all blacks feel the same way. Some blacks see any threat to the perpetuation of the race, including abortion, as an act of genocide. Maybe her black co-worker was reacting, or over-reacting, to our fear of extinction.

A female caller who telephoned a radio station in Washington, D.C., when I appeared on a morning talk show program there wanted to know if I felt there were racial differences in sneezing.

"Sneezing?" I asked.

"Well, whether blacks are more likely than whites to cover their mouths when they sneeze," she explained. The woman, who said she was white, said that a black man had cursed at her when she sneezed on a crowded bus.

"Well," I responded, "maybe the man saw her color and immediately felt it was the 'white entitlement' I wrote about. But most people would view her actions as rude or inconsiderate. Not racial in any way. We've been taught to cover our mouths when we sneeze. People who don't are viewed as lacking basic home training. And that applies to blacks and whites."

A black woman in her thirties who works as an interior designer said she read this book and wanted to tell me about something that had happened to her. She received her bachelor's degree from Rutgers University in New Jersey in the early 1990s, had worked at two major design companies in New York, and now runs her own business in New Jersey. In 1999, her company was being considered for a major contract. During an interview with two white men at the firm offering the work, she was grilled not only about her professional background but also about her educational background. When she told them that she attended Rutgers, they asked her what her SAT scores were.

"Can you believe that?" she asked me. "My SAT scores. How many professional people out there who own their own companies are asked what their SAT scores were?"

Of the many things I wrote about in the book, the one singular perception that provoked the most discussion and debate was—drum-roll here—"the hair thing!" Go figure. At every stop on the book tour, I was questioned, challenged, probed, and queried about why blacks would take offense to a white woman shaking her hair.

"I don't understand why black women would get so upset about that" I was questioned countless times.

"Black women shake their hair?" said one white radio host in Cleveland. "Should white women get upset about it?"

Deep down, I think most whites get the hair thing. They may be reluctant to admit it, but several told me so. Jacques Lowe, the noted Kennedy photographer, who died this year, said he became absolutely livid when a young white woman at a restaurant kept flicking her hair over her dinner.

"I finally told her to stop it or I'd get the maitre d' to have her table changed," said Mr. Lowe.

The young, blonde, female personal escort who accompanied me around Virginia during my book tour suggested that the hair thing is often about insecurity. "When I attended Stanford, I was on the swimming team and almost made the Olympic team," said the woman who wore her hair in a sassy short cut. "My roommate had shoulder-length blonde hair. Whenever we went out, socially, to a party, dinner, or an event on campus, she would spend an hour washing, blow-drying, and preening her hair. She would walk into a room and immediately start flipping it.

"You know what I think," she went on. "I was known around campus because I was on the swim team and other students looked up to me because of it. I think she was insecure because she didn't have something special or different other than her long blonde hair."

An interesting side note. My book-tour escort also told me that she too had been subjected to presumptions based on stereotypes. Whenever she told people that she attended Stanford and was on the university's swim team, they automatically assumed she was at the school on an athletic scholarship.

"Actually, I was there on an academic scholarship," she said.

A white woman who wrote to me from West Virginia said the hair thing was also about power. The woman said she grew up poor, in a rural section of the state. She described herself as having "dirty blonde" hair that flowed down below her waist. She learned at an early age that her hair made her special in some ways.

"The black kids who live in town always commented on my long hair," she wrote. "At school, they would ask if they could comb it. Sometimes, they'd just reach over and touch it. It made me feel pretty and special. Something I didn't always feel growing up."

Lorraine Toussaint, the actress who stars on *Any Day Now,* the popular Lifetime television series with Amy Potts, said during an appearance on a morning Los Angeles television program with me that when she wore her hair in dreadlocks, she was always cast as a prostitute, drug addict, or single mother on welfare. After she straightened her hair, she got her first on-screen kiss. She now plays a lawyer.

The hair thing is real, not imagined.

Several whites complained that while the hardcover was subtitled *The Everyday Interactions That Get under the Skin of Blacks and Whites,* the majority of experiences focused on blacks, not whites.

Unfortunately, this is true.

I had hoped to get white Americans to open up and bear their souls about race and race relations in America. I gave the whites I interviewed or tried to interview personal assurance of anonymity, if that's what it took to solicit their honest opinions. I offered to read back quotes and comments to them before going to final print. I asked white friends to vouch for me with their white friends in the hope of finding a broader cross-section of whites. I sent press kits containing my *Times* biography, a copy of the *Times* story that lead to the book, and a questionnaire to prospective focus groups so that participants would know something about me, my background, and my mindset. And still, most whites refused to talk. Those who did were often monosyllabic or felt the need to be politically correct. Some spoke under the cloak of anonymity but wanted to have the final say on their pseudoidentities.

"Whites do not have the vast experiences with race that many blacks have, so it is difficult for them to talk about racial prejudice without appearing naïve and uninformed," Dr. Alvin Poussaint, a professor of psychiatry at Harvard, told me during a chance meeting in Los Angeles where he was promoting his book on suicide. Ms. Vargas told me that whites in a focus group that appeared on the *20/20* segment expressed reluctance to talk freely because they feared they would be labeled racist if they said the wrong thing on a national television program.

"Maybe that should be your next book," said one white man who attended a book signing in Chicago. "Why Whites Won't Talk about Race." He smiled at his suggestion. He'd heard me on a morning radio show and decided to come to the bookstore to meet me in person. He said that middle-aged white men felt that race matters were "no-win situations."

"If you do talk about race and say the wrong thing, there's Hell to pay," he said. "If you keep quiet, then blacks say you don't want to talk about racial matters. You're damned if you do; damned if you don't, if you know what I mean."

There were those who criticized me for failing to offer solutions to America's racial divide. Believe me, if I had the answers, I'd go tell it on the mountain, over the hills, and everywhere. But I don't. It's not like they're contained in some moldy cheese in my refrigerator and I'm just waiting for the right patent to come along.

What I aspired to do in writing the book was to get people talking among themselves, families, coworkers and friends about the things we do and say behind closed doors, about the reasons why these centuries-old stereotypes still exist, and about what we as individuals can do about race beyond blaming institutions.

So the next time a white woman flings her hair in my face, maybe I'll think she's trying to flirt with my brother-in-law, not humiliate him. Maybe, the next time a young black man approaches a lone white woman on an elevator, she will hold the door for him, not let it close in his face. Maybe, neither blacks nor whites will be so quick to take liberties by addressing people they do not know by their first names. Maybe we should talk to each other, not at or about each other. And maybe, we all shouldn't be so thin-skinned.

Lena Williams
SEPTEMBER 2001

ACKNOWLEDGMENTS

IT WOULD take several pages of this book to thank all of the friends, colleagues, and the very special individuals who played a role in bringing this book to life. Those listed here are just a few who contributed in many ways, small and large, by lending an ear, an eye, or moment of their time. I couldn't have gotten through these past two years without you.

First and foremost to my editor, Jane Isay, for believing in me from the start and for being not only my editor (and a darn good one I might add!) but also my friend. Special thanks to David Smith, the *New York Times* editor who planted the idea that led to the story that led to the book: none of this would have been possible without your prodding and skillful eye. To John Ekizian, an agent and a friend, thanks for teaching me the literary business.

To the rest of the gang: Michael Anderson, for cajoling and consoling me throughout this project and for all your constructive criticism; the Rev. Michael Beckwith, for your spiritual inspiration; Jennifer Bristol, who retrieved 25,000 lost words from cyberspace; Joel Brokaw, for all the long-distance pick-me-ups; Diane Camper; Paul Delaney; the Rev. A. Michael Charles Durant, who gave me the wisdom to let God be my guide; Eleanor Farrar; Jill Gerston; Beverly Jackson; Shawn Kennedy; Barry Lipton, for being a mentor and a friend and for turning over your computer so I could finish the book; Jacques Lowe; Dr. Lorna McFarland; Francine McDuffie; Kenneth Noble; Ronald Prince, who's been part of this project from the time it was only a news story; Pauline Schneider; Judy Twersky; Isabel Wilkerson; and Terrie Williams for all those inspirational notes.

To the folks at Harcourt with special kudos to David Hough (thanks for a great edit and a great sense of humor); Marissa Del Fierro; Dan Farley; Jennifer Holiday; and Schuyler Huntoon.

Last and definitely not least to my beloved family: Ralph Williams; Gloria Grinage; Ada Williams; Barbara Williams Turner; Ronald Williams; Francis Grinage Sr.; Joanne Williams; Carl Turner; Clarice Williams; Mark Williams; Frank Grinage Jr.; Antoinette Williams Lynch; Dennis Lynch; Darren Grinage; Angelique Williams; Lauren Williams; Yorel, Karen and Brittany (the great nieces); Deriek Williams, and to my late brother, John Tyrone Williams, a firefighter who died in the line of duty—you are forever in my heart. We've done our parents proud. May God bless and keep you.

To all of the above, my deepest and most heartfelt thanks. Any errors or oversights contained within these pages are borne by the messenger.